Casting and Forming of Advanced Aluminum Alloys

Casting and Forming of Advanced Aluminum Alloys

Editor

Hélder Puga

MDPI • Basel • Beijing • Wuhan • Barcelona • Belgrade • Manchester • Tokyo • Cluj • Tianjin

Editor
Hélder Puga
University of Minho
Guimarães
Portugal

Editorial Office
MDPI
St. Alban-Anlage 66
4052 Basel, Switzerland

This is a reprint of articles from the Special Issue published online in the open access journal *Metals* (ISSN 2075-4701) (available at: https://www.mdpi.com/journal/metals/special_issues/casting_forming_aluminum_alloys).

For citation purposes, cite each article independently as indicated on the article page online and as indicated below:

LastName, A.A.; LastName, B.B.; LastName, C.C. Article Title. *Journal Name* **Year**, *Volume Number*, Page Range.

ISBN 978-3-0365-8146-0 (Hbk)
ISBN 978-3-0365-8147-7 (PDF)

Cover image courtesy of Hélder Puga

Contents

About the Editor

Hélder Puga

Hélder Puga, Assistant Professor at the Department of Mechanical Engineering at the School of Engineering of the University of Minho, focuses his research on advanced manufacturing and processing. His research is devoted to designing ultrasonic tools/equipment and techniques for added value in intelligent manufacturing, including methods to aid in the sustainable product design/manufacturing of automotive, aeronautic, and biomedical devices. He has co-authored 98 papers, with an h-index of 24.

Preface to "Casting and Forming of Advanced Aluminum Alloys"

The automotive and aeronautical industries have faced mounting pressure to reduce gas emissions and meet consumer expectations in recent years. As a result of these challenges, aluminum alloy casting has undergone a transformative wave of innovation. Thus, smaller and more efficient aluminum alloy casting components are emerging with equivalent or even better mechanical properties. These advancements have not only contributed to enhanced fuel economy and reduced gas emissions, but have also heralded an era of sustainable manufacturing practices.

In the production of well-formed and reliable metal components that align with customers' needs, aluminum alloy metal casting has proven to be the most cost-effective method. As a result, light alloys, particularly Al-based alloys, are increasingly used in structural component manufacturing. These developments have propelled the casting and forming of advanced aluminum alloys to the forefront, requiring substantial progress in processing techniques.

Combining low density, high corrosion resistance and strength, workability, and excellent electrical and heat conductivity, these materials demand cutting-edge processing techniques. Research and innovation in aluminum alloy applications has the power to significantly improve people's lives and provide significant support to manufacturing.

Future generations must embrace sustainability principles and expand the scope of what is achievable in line with the desires of both individuals and the industrial sectors. Our goal is to attract the attention of the scientific and industrial communities by offering a useful resource that showcases the latest advances in aluminum alloy research. It will foster collaboration and provide a forum of knowledge and inspiration for researchers, engineers, and professionals alike. Our collective future will more sustainable and technologically advanced if we are all informed about these developments.

Together, we can forge a path towards a more sustainable, innovative, and prosperous future.

Hélder Puga
Editor

metals

MDPI

Editorial

Casting and Forming of Advanced Aluminum Alloys

Hélder Puga

CMEMS-UMinho, Center for Microelectromechanical Systems, Mechanical Department, University of Minho, 4710-057 Braga, Portugal; puga@dem.uminho.pt; Tel.: +351-252-510-220

Received: 6 April 2020; Accepted: 8 April 2020; Published: 9 April 2020

1. Introduction and Scope

The automotive and aeronautical industry´s response to the environmental impact provoked by gas emissions and consumer expectations has driven aluminum alloy casting changes in recent years. Light weighting and downsizing have led to the production of smaller and more efficient aluminum alloy casting components with the same or improved mechanical properties, helping fuel economy and gas emissions. Aluminum alloy metal casting is the most cost-effective method to produce shaped and soundness metal components able to answer to clients' requirements. As a result, an increasing trend of using light alloys, namely Al-based alloys, for the production of structural components, has been undergoing major developments. In recent years, we have been confronted with new developments in casting and forming of advanced aluminum alloys, namely in demand for new processing techniques.

The Special Issue *Casting and Forming of Advanced Aluminum Alloys* provides a theoretical and practical understanding of the metallurgic principles in the casting process, advanced melt treatment techniques, and forming process, aiming to enhance the mechanical performance of the aluminum alloys further.

Breakthrough innovations are needed to boost the quality of aluminum alloy research as well as to address the many challenges faced by the need to develop new, advanced techniques in processing enhancing the combination of characteristics such as the low density, the high corrosion resistance, high strength, workability and high electrical and heat conductivity. Contributing to the research into, and innovation of, aluminum alloy application will improve the lives of people. A reflection on the future should be inspired by the goals of sustainability to stretch the boundaries of what is feasible in the function of what is desirable for the people and manufacturing industries. We hope that this Special Issue can deserve the attention of the scientific and industrial community to keep up-to-date with the latest developments in aluminum alloy research.

2. Contributions

Conventional aluminum alloys have reached their limits concerning their mechanical properties. Thus, finding effective and reliable solutions to develop new alloys as well as new technologies for its processing remains a demanding challenge. In this Special Issue of *Metals*, ten articles have been published, covering a wide scope of recent progress and developments regarding some aspects of aluminum alloys, including their processing, microstructure and mechanical properties, corrosion and surface quality. These aspects can be combined into two main groups (1) casting and (2) forming an aluminum alloy.

2.1. Casting Aluminum Alloy

The quality and competitiveness of a casting strongly depend on the quality of the molten alloy and the technology used to produce it. Aluminum alloy casting is not an easy process, since these alloys are prone to dendritic and heterogeneous structures, as well as the absorption of hydrogen during melting. Thus, a specific melt processing operation is required in order to reduce and control the level of porosity in the microstructure after solidification. Controlling the microstructure of aluminum alloys is of primary importance to achieve high mechanical performance, requiring suitable degassing, modification, and refinement techniques. Conventional casting is a well-established process for the manufacture of a wide variety of aluminum components. Nevertheless, achievable casting performance is limited due to defects that emerge during melt processing and solidification. In recent years, an effort has been made to develop new and reliable techniques to control the microstructure of several engineering alloys, with a particular emphasis on Al-based alloys, to overcome the problems associated with traditional melt techniques.

The effect of traveling magnetic fields (TMFs) on the grain and micro-pore formation in an Al-Cu alloy was studied by Xu et al. [1]. In this study, it was reported that the forced convection induced by TMF break the dendrites, refine the grain size, and promote liquid feeding, leading to a decrease in the volume fraction of the porosity and improved mechanical property. The microstructural evaluation and corrosion resistance of a semisolid Cast A356 alloy were studied by Gebril et al. [2]. For that, a combination of *as-cast* and semisolid casting using a cooling slope processed by equal channel angular pressing (ECAP) was used. Eskin and Wang [3], Kudryashova et al. [4] and Puga et al. [5] applied the ultrasonic vibration to study the effect of ultrasonic melt treatment in the solidified microstructure of aluminum alloys. The role of the roll-separating force in the high-speed twin-roll casting of aluminum alloys was examined by Kim et al. [6]. For that, a traditional twin-roll casting (TRC) process was designed to combine metal casting and hot rolling into a single operation.

2.2. *Forming Aluminum Alloy*

The forming behavior of aluminum alloys has been assessed through the application of different processes, which include extrusion, stretching, bending and hydroforming.

Ciuffini et al. [7] reported and discussed the relationship between the surface quality and the use of internal liquid nitrogen cooling during the aluminum extrusion. Li et al. [8] focused attention on the mechanisms of cavity nucleation and cavity growth of a 5A70 aluminum alloy during superplastic deformation. The results demonstrated a clear transition from diffusion growth to superplastic diffusion growth and plastic-controlled growth at a cavity radius larger than 1.52 and 13.90 μm. On the other hand, for the same class of aluminum alloy, Li et. al. [9] verified that the superplastic behavior depends on temperatures, strain rates, and precipitated phases during superplastic deformation. Yang et al. [10] outline the findings of the study of the effects of prebending radii on the hardness, tensile strength, yield strength, elongation and HCF performance of the 7075 aluminum alloy after creep age forming.

3. Conclusions and Outlook

The present Special Issue emerges from recent developments in casting and forming, highlighting different aspects of the processing, metallurgical and mechanical behavior of aluminum alloys. All the contributions outline problems and give solutions to achieve further progress in advanced aluminum alloy applications.

As Guest Editor of this Special Issue, I would like to acknowledge the contribution of all who kindly submitted their articles, all the reviewers for their efforts in ensuring a high-quality publication, and those who want to share their work. Finally, it is a great pleasure to acknowledge the professional support by the *Metals* Team, particularly Betty Jin, and her help and support.

Conflicts of Interest: The author declares no conflict of interest.

References

1. Xu, Y.; Wei, L.; Han, B.; Guo, E.; Wang, M.; Su, Y. Effect of a Traveling Magnetic Field on Micropore Formation in Al-Cu Alloys. *Metals* **2018**, *8*, 448. [CrossRef]
2. Gebril, A.M.; Omar, Z.M.; Mohamed, F.I.; Othman, K.N. Microstructural Evaluation and Corrosion Resistance of Semisolid Cast A356 Alloy Processed by Equal Channel Angular Pressing. *Metals* **2019**, *9*, 303. [CrossRef]

3. Eskin, D.; Wang, F. Joint Effect of Ultrasonic Vibrations and Solid Metal Addition on the Grain Refinement of an Aluminium Alloy. *Metals* **2019**, *9*, 161. [CrossRef]
4. Kudryashova, O.; Khmeleva, M.; Danilov, P.; Dammer, V.; Vorozhtsov, A.; Eskin, D. Optimizing the Conditions of Metal Solidification with Vibration. *Metals* **2019**, *9*, 366. [CrossRef]
5. Puga, H.; Barbosa, J.; Carneiro, H.V. The Role of Acoustic Pressure during Solidification of AlSi7Mg Alloy in Sand Mold Casting. *Metals* **2019**, *9*, 490. [CrossRef]
6. Kim, M.; Kim, H.; Kim, S.; Kumai, S. Role of Roll Separating Force in High-Speed Twin-Roll Casting of Aluminum Alloys. *Metals* **2019**, *9*, 645. [CrossRef]
7. Ciuffini, F.A.; Barella, S.; Cecca, D.C.; Gruttadauria, A.; Mapelli, C.; Merello, L.; Mainetti, G.; Bertoletti, M. Surface Quality Improvement of AA6060 Aluminum Extruded Components through Liquid Nitrogen Mold Cooling. *Metals* **2018**, *8*, 409. [CrossRef]
8. Li, S.; Jin, S.; Huang, X. Cavity Behavior of Fine-Grained 5A70 Aluminum Alloy during Superplastic Formation. *Metals* **2018**, *8*, 1065. [CrossRef]
9. Li, S.; Huang, X.; Jin, S. Superplastic Behavioral Characteristics of Fine-Grained 5A70 Aluminum Alloy. *Metals* **2019**, *9*, 62. [CrossRef]
10. Yang, D.; Miao, J.; Zhang, F.; Fu, Z.; Liu, Y. Effects of Prebending Radii on Microstructure and Fatigue Performance of Al-Zn-Mg-Cu Aluminum Alloy after Creep Age Forming. *Metals* **2019**, *9*, 630. [CrossRef]

metals

MDPI

Article

The Role of Acoustic Pressure during Solidification of AlSi7Mg Alloy in Sand Mold Casting

H. Puga [1,*], J. Barbosa [1] and V. H. Carneiro [2]

1 CMEMS—UMinho, Campus de Azurém, 4800-058 Guimarães, Portugal; kim@dem.uminho.pt
2 MEtriCS—UMinho, Campus de Azurém, 4800-058 Guimarães, Portugal; vitorhcarneiro@hotmail.com
* Correspondence: puga@dem.uminho.pt; Tel.: +351-253-510220

Received: 25 March 2019; Accepted: 25 April 2019; Published: 27 April 2019

Abstract: New alloy processes have been developed and casting techniques are continuously evolving. Such constant development implies a consequent development and optimization of melt processing and treatment. The present work proposes a method for studying the influence of acoustic pressure in the overall refinement of sand cast aluminum alloys, using and correlating experimental and numerical approaches. It is shown that the refinement/modification of the α-Al matrix is a consequence of the acoustic activation caused in the liquid metal directly below the face of the acoustic radiator. Near the feeder, there is a clear homogeneity in the morphology of the α-Al with respect to grain size and grain circularity. However, the damping of acoustic pressure as the melt is moved away from the feeder increases and the influence of ultrasound is reduced, even though the higher cooling rate seems to compensate for this effect.

Keywords: ultrasonic melt refinement; sand casting: acoustic radiator; α-Al grain size; aluminum alloy

---

1. Introduction

Aluminum alloys are thriving in the automotive, aeronautics and aerospace industries, displacing applications that have traditionally been occupied by other alloys [1]. The performance of a mechanical component is often conditioned and limited by the characteristics of the materials themselves, as well as by limitations of the manufacturing processes and, more specifically, by their microstructure. Although, their use is widespread, the casting of aluminum alloys is not an easy process, since they are prone to nucleate and grow coarse and dendritic microstructures [2,3]. Additionally, aluminum alloys are characterized by a high absorption of hydrogen during melting and casting [4,5]. Thus, the increasing use of aluminum components with superior mechanical and fatigue properties requires suitable and high-efficiency casting processes [6]. This includes the melt treatment [7] to develop suitable microstructures, eliminate inclusions, and reduce porosities and shrinkage defects, which are the main cause of failure in aluminum components [8].

In the industrial practice of aluminum casting, three melt treatment operations are usually carried out in addition to the removal of slag, all of them chemically-based and presenting significant environmental impacts: (i) degassing, by reducing the hydrogen content of the melt, which is achieved by gas purging using mainly inert gases [9,10]; (ii) microstructure refinement, by the addition of Al-Ti-B type master alloys in proportions adjusted to the specific aluminum alloy [11–13]; (iii) eutectic silicon modification, usually carried out by the addition of master alloys containing Sr [12,14]. Although these three stages (degassing, grain refinement, modification of eutectic silicon), are vital to improving the mechanical performance of castings, the need to find more efficient, viable and clean treatment alternatives has fostered the search for new melt treatment techniques and technologies.

In the last decade, we have witnessed the development of highly efficient aluminum melt treatment techniques based on the use of acoustic energy [15–19]. The influence of ultrasound on the refinement/modification of the microstructure is based on physical phenomena due to the high acoustic intensity propagated through the liquid metal [17]. Two mechanisms have been proposed to explain the refinement of microstructures by ultrasound, dendritic fragmentation and cavitation induced heterogeneous nucleation [20,21]. However, the mechanism of heterogeneous nucleation induced by cavitation seems to be the most valid hypothesis, being supported by the majority of researchers who have worked in this field [22]. However, in order for this technique to be efficient, it is imperative that the ultrasonic system is correctly designed according to the specific needs of the casting process. This enhances the overall success of the technique and optimizes the casting component integrity.

For that purpose, the present work aimed to study the interaction between the requirements imposed by the melt conditions (i.e., the melt temperature/volume) and the constraints imposed by the manufacturing process (geometry of casting) in the optimization of an ultrasonic system and its impact on the overall microstructures [23]. Furthermore, considering their physical processing, a numerical model was used to investigate the associated acoustic pressure fields developed in the transmission medium and their role in the grain refinement.

2. Methodology

2.1. Experimental Setup and Procedure

The excellent castability and ductility, combined with a reduced tendency to defect formation, have been crucial factors in increasing the application of AlSi7Mg based alloys in the production of structural castings cast by sand molding [24,25]. The mechanical properties of such castings can be improved by solubilizing and aging treatments, providing uniform distribution of Mg and Si precipitates in the aluminum matrix [26,27].

The melt charges used in this work were prepared from an AlSi7Mg0.3 alloy, with the composition presented in Table 1.

Table 1. Chemical composition of standard AlSi7Mg0.3 alloy.

Alloy	Chemical Composition (wt%)									Source
	Si	Fe	Mg	Cu	Mn	Zn	Ti	Al	Res.	
Std AlSi7Mg0.3	6.5–7.5	0.6	0.20–0.45	0.25	0.35	0.35	0.25	Bal.	0.15	(1)
Used alloy	7.44	0.0	0.32	0.07	0.07	0.05	0.11	91.53	0.21	(2)

(1) According to Aluminum Association, Inc.
(2) Composition of the alloy used in the experimental work.

Melting charges weighing 10 kg were previously cut from a primary melt commercial ingot and later washed and dried to eliminate the cutting lubricant. After melting and overheating at the pre-established testing temperature (720 ± 5 °C), the melt was kept isothermal for a 30 min period for better homogenization. After this period, the melt was degassed for 5 min using ultrasound technology. The 5-minute period and the operational parameters were chosen according to works previously published by the authors [28]. After degassing, the melt was allowed to cool and was poured at 700 ± 5 °C into a sand mold, as shown in Figure 1. Immediately after pouring, the pre-heated acoustic radiator was positioned over the feeder with an immersion depth of 15 mm in the melt, to ensure the maximum transfer of acoustic energy to the melt from the thick zone (feeder) to the thin areas of the casting. Acoustic energy was supplied until the metal reached the solidus temperature +10 °C.

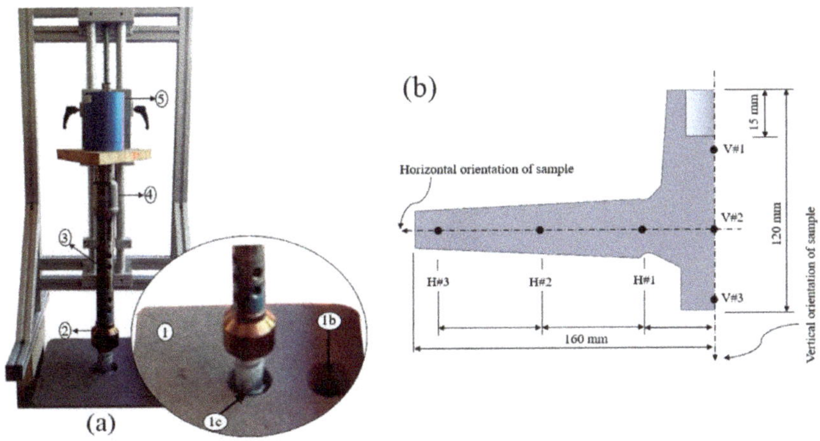

Figure 1. (**a**) Experimental setup: (1) sand mold, (1b) pouring basin, (1c) feeder, (2) acoustic radiator, (3) waveguide, (4) booster, (5) transducer 20 kHz; (**b**) Geometric model where V#1 to V#3 correspond to the positions for sample characterization (Note: mirrored symmetry).

Samples for microstructure characterization were taken from every cast sample by sectioning them perpendicularly to their vertical and horizontal axis (respectively, V#1 to V#3 and H#1 to H#3), according to Figure 1b. They were ground using 1200 μm SiC and polished down to 1 μm. An optical microscope (OM) (LEICA DM 2500M) and the ImageJ v1.46 computer application were used to determine the average grain size, d, and circularity, R_n, using Equations (1) and (2) where A corresponds to the area and P to the perimeter of the α-grains.

$$d = 2 \times \sqrt{\frac{A}{\pi}} \tag{1}$$

$$R_n = \frac{4 \times \pi \times A}{P^2} \tag{2}$$

2.2. Computational Modeling

In order to study the acoustic propagation in liquid metal, a simulation model was developed using the COMSOL v5.2a Multiphysics module—Acoustic Piezoeloectric Interaction, Frequency Domain, according to Figure 2. Considering that the acoustic wave propagation is linear, and the shear stresses are negligible for fluids, the acoustic pressure can be calculated by applying the following wave equation [6,7]:

$$\nabla\left(\frac{1}{\rho}\nabla P\right) - \frac{1}{\rho c^2}\frac{\partial^2 P}{\partial t^2} = 0 \tag{3}$$

where ρ is the density of the liquid metal, c is the sound velocity in the liquid metal and t is time. For the case of a harmonic wave in time, the pressure varies according to:

$$P(r,t) = p(r)^{i\omega t} \tag{4}$$

where, $\omega = 2\pi f$ is the angular frequency and p is the acoustic pressure. Assuming that the same harmonic is time dependent, in the same terms, Equation (3) can be reduced, through Equation (4), to the Helmholtz equation:

$$\nabla\left(\frac{1}{\rho}\nabla p\right) - \frac{\omega^2}{\rho c^2}p = 0 \tag{5}$$

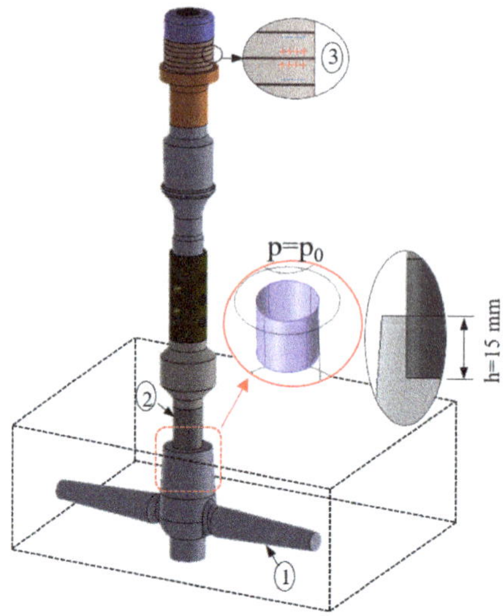

Figure 2. Geometry modeled using COMSOL Multiphysics—Acoustic Piezoloectric (PZT) Interaction, Frequency Domain. (1) acoustic medium, (2) Ti6Al4V acoustic radiator, (3) Piezoloectric (PZT) polarization.

The Acoustic Piezoloectric Interaction module in COMSOL Multiphysics was used to perform an analysis in the frequency domain. This combines the effects of (i) sound pressure and (ii) piezoelectric, linking the variations of acoustic pressure with the solids that are actuated by the piezoelectric effect of PZT. The physical interface also includes electrostatic elements to solve the electric field in the piezoelectric material. The Helmholtz equation is solved in the fluid domain and the structural equations in the solid domain, together with the constructive relations necessary for the piezoelectric modeling. The physical interface that solves the Helmholtz equation is suitable for the present study, in the domain of linear frequencies with harmonic variation of the pressure field.

In order to evaluate the profile of the acoustic pressure during the deformation of the solids actuated by the PZT piezoelectric effect, water was used since this is a suitable liquid medium to simulate the refinement / modification mechanism that occurs in melts of aluminum alloys at 660–700 °C [29]. With the appropriate boundary conditions, the Helmholtz equation can be solved through a range of numerical methods [6,8]. The accuracy of the numerical solution of the Helmholtz equation depends significantly on the wave number $k (k = \omega/c)$. The main boundary conditions are described as, (i) $p = 0$ (condition of total waves reflection, attributed to the liquid-air interfaces); (ii) $p = p_0$ (interface of the acoustic radiator with the liquid metal); (iii) $\delta p/\delta n = 0$ (condition of "rigid walls" attributed to the lateral walls of the acoustic radiator); (iv) $(1/\rho)(\delta p/\delta n) + i\omega p/Z = 0$ (acoustic impedance limit condition Z attributed to the container walls).

3. Results and Discussion

Aluminum alloys are known for their high affinity to hydrogen at high temperature, promoting the appearance of numerous inclusions in the microstructure after solidification, with porosity being the most relevant and harmful defect compromising the strength of the component [4,30,31]. To assess the similarity of the porosity levels in samples, the melts were degassed before pouring at the desired temperature (700 ± 5 °C [31]), using the Multi-Mode Frequency, Modulated (MMM) ultrasonic technology, to ensure that the density of the bath was identical for every experiment. The average density reached in the experiments was 2.68 ± 0.1 (0.5% ± 0.07 porosity).

Figure 3 shows the variation of the α-Al grain size and circularity of AlSi7Mg0.3 alloy processed by ultrasound, with an average frequency of 19.9 ± 0.2 kHz and 40% of the system power

(400 W—corresponding to approximately 60 μm), evaluated in a longitudinal section of the feeder (V#1 to V#3 samples, according to Figure 1). The respective microstructures are shown in Figure 4a–c.

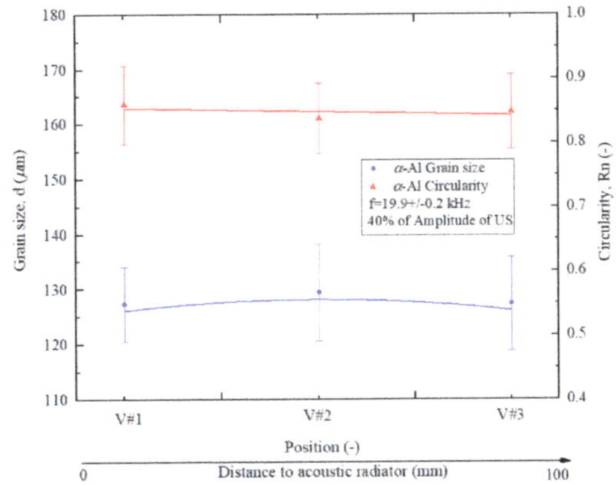

Figure 3. Variation of the α-Al grain size and circularity with the distance to the acoustic radiator in the feeder.

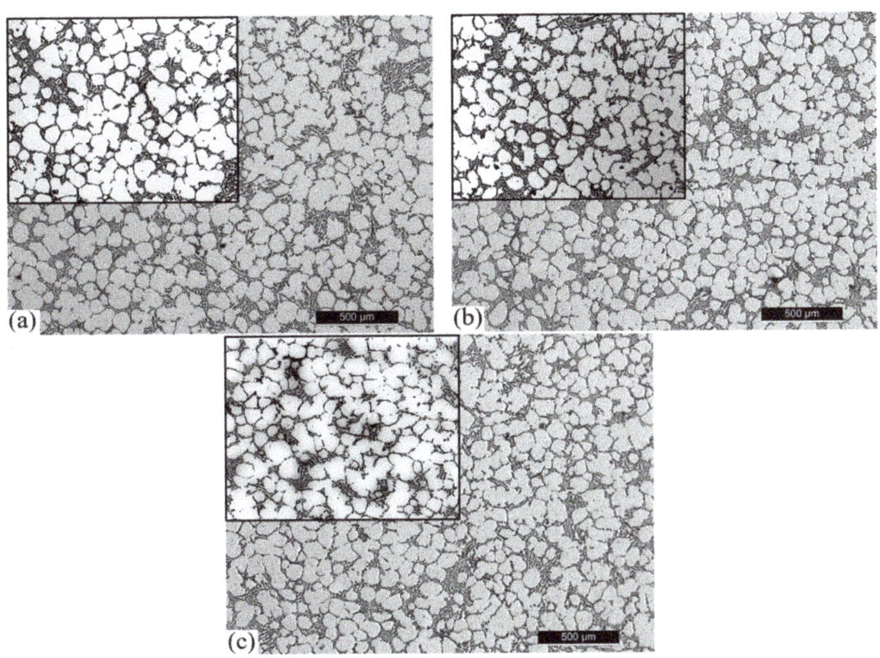

Figure 4. Examples of microstructures of AlSi7Mg0.3 alloy processed by ultrasound, at 19.9 ± 0.2 kHz average frequency evaluated in a vertical section of the feeder: (**a**) V#1; (**b**) V#2 and (**c**) V#3 samples, according to Figure 1.

The experimental results suggest that the feeder (central zone of the feeder and directly below the acoustic radiator) presents a microstructure with a well-defined globular morphology. Overall, the variation of grain size and circularity apparent was minimal, as can confirmed by the standard deviations presented in Figure 3. Indeed, the average grain size and circularity for all characterized positions were, respectively, 120 μm and 0.8. As demonstrated by Puga et al. [32] and Eskin et al. [29] the zone below the flat acoustic radiator is able to promote extremely intense cavitation, being able to promote nucleation and further to this, an increase in the quantity of α-Al grains, and consequently, the reduction of their diameter and a globular morphology.

Figure 5 shows the variation of the α-Al grain size and circularity of AlSi7Mg0.3 alloy processed by ultrasound, at the average frequency of 19.9 ± 0.2 kHz (resonance frequency at high temperature), with the distance to the acoustic radiator in the feeder (H#1 to H#3 samples, according to Figure 1). The respective microstructures are shown in Figure 6a–c.

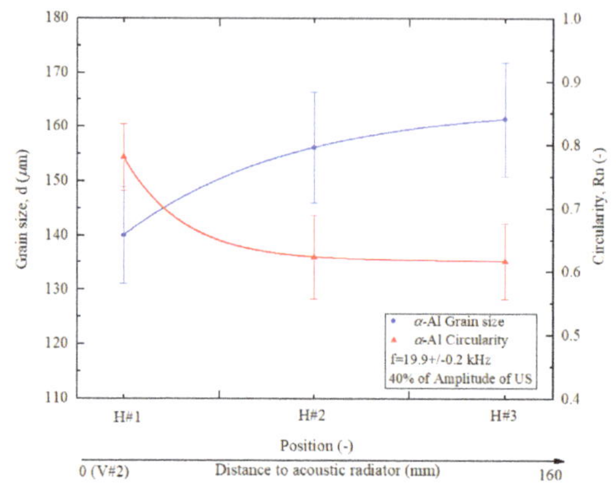

Figure 5. Variation of the α-Al grain size and circularity with the distance to the acoustic radiator in the feeder.

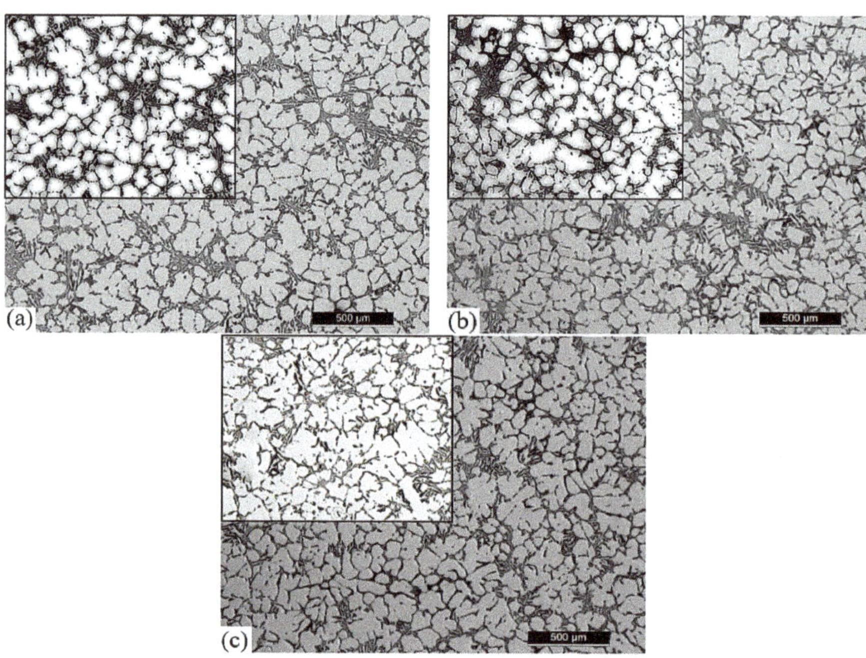

Figure 6. Examples of microstructures of AlSi7Mg0.3 alloy processed by ultrasound, at 19.9 ± 0.2 kHz average frequency evaluated in a horizontal section of the feeder: (**a**) H#1, (**b**) H#2 and (**c**) H#3 samples, according to Figure 1.

The results presented suggest that with increasing distance to the acoustic radiator the α-Al grain size tends to increase, and the grain circularity tends to decrease. Although the grain size tends to increase, the matrix continues to present a quasi-globular morphology with some indications of rosette grains. This is an opposite tendency compared to what happens with the traditional processes of sand casting, where the matrix tends to present a dendritic morphology, even after chemical melt treatments.

Furthermore, the results presented in Figures 5 and 6 suggest that the cooling rate influences the grain morphology. Although the effect of ultrasound promotes a globular matrix, the fast cooling rates in thinner sections (e.g., position H#3 with a section solidification module 0.41) tends to overlap the effect of ultrasound. However, this effect also allows the formation of a thinner but non-dendritic microstructure, which is beneficial for the mechanical properties. However, in sections directly affected by the acoustic radiator (e.g., feeder—Figures 3 and 4), it is suggested that the morphology of the Al matrix (grain size and circularity) may be directly affected by the cavitation mechanism.

Contrary to the traditional process of refinement, where the addition of the master alloy Al-Ti-B is generally used, in the present approach this issue is not considered. Overall, the applied acoustic intensity is sufficient for: (i) converting the displacement of piezoelectric into kinetic energy in the liquid; (ii) creating acoustic cavitation and acoustic streams able to promote the nucleation and homogenization in the melt; (iii) increasing the temperature of the medium due to bubble collapse, which can promote and accelerate the α-Al grains.

A numerical simulation study was performed to validate the aforementioned hypothesis and experimental results. Figure 7 presents the numerical results of the axial displacement (evaluated in solid parts and the acoustic pressure in the melt), for the whole system, when the acoustic radiator was immersed in water at a depth of 15 mm, according to the boundary conditions represented in Figure 2.

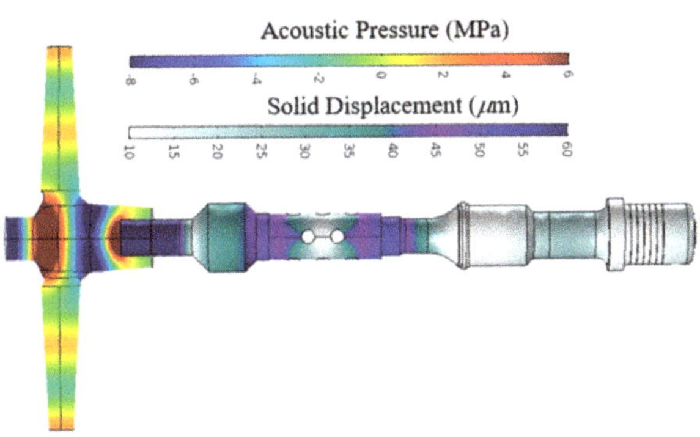

Figure 7. Numerical results of the solid displacement and acoustic pressure obtained for the ultrasonic system apparatus.

According to the eigenfrequency results of the acoustic radiator, the first longitudinal compression mode was located at 20.23 kHz. This matches well with the operation frequency of the transducer, as can be proved by the experimental electrical impedance results measured by the authors. Thus, in order to numerically quantify the acoustic pressure in the medium, a domain frequency study was performed at f_0 = 20.20 kHz. Figure 8a,b presents the numerical values of acoustic pressure along the horizontal section with the distance to the acoustic radiator in the feeder (H#1 to H#2) and the vertical component beneath the feeder (V#1 to V#3).

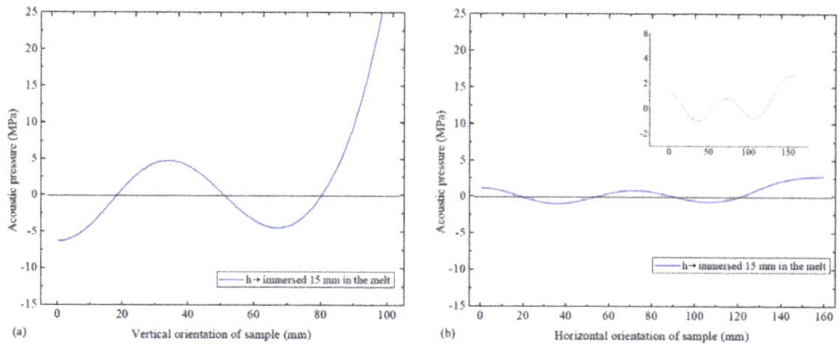

Figure 8. Numerical results of the acoustic pressure obtained in the (**a**) vertical and (**b**) horizontal directions.

As can be observed in Figure 8a, a sinusoidal curve with a positive and negative acoustic pressure of at least 5 MPa in the vertical component is reached in the melt. According to Eskin's work [29] the threshold of acoustic pressure to verify cavitation is around 2 MPa. Indeed, according to experimental results for parameters numerically evaluated, the level of cavitation and acoustic streams at 40% system power (using the same acoustic radiator that numerically simulated) is totally defined, as can be observed in Figure 9. It may be clearly observed that in corresponding conditions (i.e., similar fluid and boundary conditions) that are initially resting (Figure 9a), it is possible to promote streaming and cavitation effects by the use of the detailed ultrasonic system.

Figure 9. Photograph of resonance cavitation field in the experimental container (400 W): (**a**) No-US, (**b**) with US activated.

Contrary to high values of acoustic pressure measured along the longitudinal section of the feeder (V#1 to V#3), in the horizontal cylindrical section (H#1 to H#2) the maximum acoustic pressure registered was 1 MPa (Figure 8b). Considering the high level of cavitation and acoustic streams created below the flat surface of the acoustic radiator, as well as the long interval time of solidification, it is suggested that a mixing and distribution of nuclei can travel along the cylindrical shape and contribute to the refining of Al grain reported in results of Figures 5 and 6. Furthermore, according to Figure 10, there is a correlation between the acoustic pressure and the grain size. It is shown that as the pressure increases, the grain size tends to decrease by an exponential decay function.

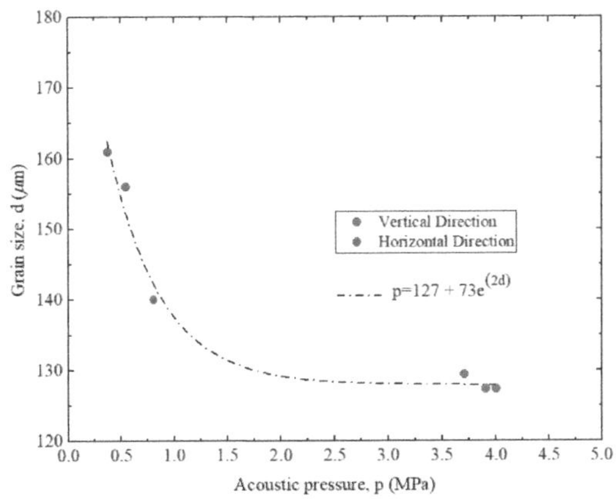

Figure 10. Effect of acoustic pressure versus grain size.

It is suggested that there is a threshold for the acoustic pressure grain refinement effect. In these particular conditions, it is apparent that for acoustic pressures higher than 2 MPa there are no significant benefits in terms of grain refinement.

4. Conclusions

This study explored the influence of acoustic pressure in the overall refinement of sand cast aluminum alloys, using experimental and numerical approaches to detail the overall pressure distribution and distance to acoustic excitation. According to the results, the following conclusions may be drawn:

(1) The mechanism of refinement/modification of the α-Al matrix is a consequence of the acoustic activation caused in the liquid metal directly below the face of the acoustic radiator and is able to be distributed by different mold cavity area branches.

(2) In areas near the feeder there is a clear homogeneity in the morphology of the α-Al with respect to grain size and grain circularity. That is, the influence of the acoustic radiator in the liquid medium immediately below its top flat face is evident.

(3) In areas that are more distant to the feeder, the acoustic pressure directly caused by the acoustic radiator tends not to induce significant alteration in the grain size due to the lower pressure; however, this is compensated for by the higher cooling rates.

(4) Knowledge of the acoustic pressure profile, together with the analysis of the positioning of the acoustic radiator for refinement/modification the α-Al matrix validated through the use of numerical models, will allow high integrity castings to be obtained, with a tendency towards increased mechanical properties when compared to traditional treatment methods.

Author Contributions: Conceptualization, V.H.C. and H.P.; Methodology, V.H.P. and H.P.; Validation, H.P. and J.B.; Formal analysis, V.H.C. and H.P.; Resources, H.P.; Data curation, V.H.P. and J.B.; Writing—original draft preparation, V.H.P. and H.P.

Funding: This research received no external funding.

Acknowledgments: This work was supported by FCT funding through the project PTDC/EMEEME/30967/2017 (NORTE-0145-FEDER-030967) and by FEDER funding through COMPETE 2020, NORTE2020, PORTUGAL2020 – Programa Operacional Competitividade e Internacionalização (POCI). Additionally, this work was supported by FCT with the reference project UID/EEA/04436/2019.

Conflicts of Interest: The authors declare no conflicts of interest.

References

1. Nguyen, R.T.; Imholte, D.D.; Rios, O.R.; Weiss, D.; Sims, Z.; Stromme, E.; McCall, S.K. Anticipating impacts of introducing aluminum-cerium alloys into the United States automotive market. *Resour. Conserv. Recycl.* **2019**, *144*, 340–349. [CrossRef]
2. Jarry, P.; Rappaz, M. Recent advances in the metallurgy of aluminium alloys. Part I: Solidification and casting. *C. R. Phys.* **2018**, *19*, 672–687. [CrossRef]
3. Liang, G.; Ali, Y.; You, G.; Zhang, M.-X. Effect of cooling rate on grain refinement of cast aluminium alloys. *Materialia* **2018**, *3*, 113–121. [CrossRef]
4. Zhang, Q.; Wang, T.; Yao, Z.; Zhu, M. Modeling of hydrogen porosity formation during solidification of dendrites and irregular eutectics in Al–Si alloys. *Materialia* **2018**, *4*, 211–220. [CrossRef]
5. Brůna, M.; Bolibruchová, D.; Pastirčák, R. Numerical Simulation of Porosity for Al Based Alloys. *Procedia Eng.* **2017**, *177*, 488–495. [CrossRef]
6. Su, H.; Toda, H.; Masunaga, R.; Shimizu, K.; Gao, H.; Sasaki, K.; Bhuiyan, M.S.; Uesugi, K.; Takeuchi, A.; Watanabe, Y. Influence of hydrogen on strain localization and fracture behavior in AlZnMgCu aluminum alloys. *Acta Mater.* **2018**, *159*, 332–343. [CrossRef]
7. Jung, J.-G.; Cho, Y.-H.; Lee, J.-M.; Kim, H.-W.; Euh, K. Designing the composition and processing route of aluminum alloys using CALPHAD: Case studies. *Calphad* **2019**, *64*, 236–247. [CrossRef]
8. Rotella, A.; Nadot, Y.; Piellard, M.; Augustin, R.; Fleuriot, M. Fatigue limit of a cast Al-Si-Mg alloy (A357-T6) with natural casting shrinkages using ASTM standard X-ray inspection. *Int. J. Fatigue* **2018**, *114*, 177–188. [CrossRef]
9. Mancilla, E.; Cruz-Méndez, W.; Garduño, I.E.; González-Rivera, C.; Ramírez-Argáez, M.A.; Ascanio, G. Comparison of the hydrodynamic performance of rotor-injector devices in a water physical model of an aluminum degassing ladle. *Chem. Eng. Res. Des.* **2017**, *118*, 158–169. [CrossRef]
10. Haghayeghi, R.; Bahai, H.; Kapranos, P. Effect of ultrasonic argon degassing on dissolved hydrogen in aluminium alloy. *Mater. Lett.* **2012**, *82*, 230–232. [CrossRef]
11. Li, J.; Huang, M.; Ma, M.; Ye, W.; Liu, D.; Sone, D.; Bai, B.; Fang, H. Performance comparison of AlTiC and AlTiB master alloys in grain refinement of commercial and high purity aluminum. *Trans. Nonferrous Met. Soc. China* **2006**, *16*, 242–253. [CrossRef]
12. Lu, L.; Dahle, A.K. Effects of combined additions of Sr and AlTiB grain refiners in hypoeutectic Al–Si foundry alloys. *Mater. Sci. Eng. A* **2006**, *435–436*, 288–296. [CrossRef]
13. Ding, W.; Xu, C.; Hou, X.; Zhao, X.; Chen, T.; Zhao, W.; Xia, T.; Qiao, J. Preparation and synthesis thermokinetics of novel Al-Ti-C-La composite master alloys. *J. Alloys Compd.* **2019**, *776*, 904–911. [CrossRef]
14. Öztürk, İ.; Hapçı Ağaoğlu, G.; Erzi, E.; Dispinar, D.; Orhan, G. Effects of strontium addition on the microstructure and corrosion behavior of A356 aluminum alloy. *J. Alloys Compd.* **2018**, *763*, 384–391. [CrossRef]
15. Barbosa, J.; Puga, H. Ultrasonic Melt Treatment of Light Alloys. *Int. J. Met.* **2019**, *13*, 180–189. [CrossRef]
16. Eskin, D.G.; Tzanakis, I.; Wang, F.; Lebon, G.S.B.; Subroto, T.; Pericleous, K.; Mi, J. Fundamental studies of ultrasonic melt processing. *Ultrason. Sonochem.* **2019**, *52*, 455–467. [CrossRef] [PubMed]
17. Tzanakis, I.; Lebon, G.S.B.; Eskin, D.G.; Pericleous, K.A. Characterisation of the ultrasonic acoustic spectrum and pressure field in aluminium melt with an advanced cavitometer. *J. Mater. Process. Technol.* **2016**, *229*, 582–586. [CrossRef]
18. Barbosa, J.; Puga, H. Ultrasonic melt processing in the low pressure investment casting of Al alloys. *J. Mater. Process. Technol.* **2017**, *244*, 150–156. [CrossRef]
19. Tuan, N.Q.; Puga, H.; Barbosa, J.; Pinto, A.M.P. Grain refinement of Al-Mg-Sc alloy by ultrasonic treatment. *Met. Mater. Int.* **2015**, *21*, 72–78. [CrossRef]
20. Kotadia, H.R.; Qian, M.; Das, A. Solidification of aluminium alloys under ultrasonication: An overview. *Trans. Indian Inst. Met.* **2018**, *71*, 2681–2686. [CrossRef]
21. Wang, G.; Wang, Q.; Easton, M.A.; Dargusch, M.S.; Qian, M.; Eskin, D.G.; StJohn, D.H. Role of ultrasonic treatment, inoculation and solute in the grain refinement of commercial purity aluminium. *Sci. Rep.* **2017**, *7*, 9729. [CrossRef] [PubMed]
22. Puga, H.; Barbosa, J.; Costa, S.; Ribeiro, S.; Pinto, A.M.P.; Prokic, M. Influence of indirect ultrasonic vibration on the microstructure and mechanical behavior of Al–Si–Cu alloy. *Mater. Sci. Eng. A* **2013**, *560*, 589–595. [CrossRef]

23. Puga, H.; Carneiro, V.; Barbosa, J.; Vieira, V. Effect of ultrasonic treatment in the static and dynamic mechanical behavior of AZ91D Mg alloy. *Metals* **2015**, *5*, 2210–2221. [CrossRef]
24. Weiss, D. Chapter 5—Advances in the Sand Casting of Aluminium Alloys. In *Fundamentals of Aluminium Metallurgy*; Lumley, R.N., Ed.; Woodhead Publishing: Sawston, UK, 2018; pp. 159–171. ISBN 978-0-08-102063-0.
25. Shangguan, H.; Kang, J.; Deng, C.; Hu, Y.; Huang, T. 3D-printed shell-truss sand mold for aluminum castings. *J. Mater. Process. Technol.* **2017**, *250*, 247–253. [CrossRef]
26. Carneiro, V.H.; Puga, H. Solution treatment enhances both static and damping properties of Al–Si–Mg alloys. *Metall. Mater. Trans. A* **2018**, *49*, 5942–5945. [CrossRef]
27. Carneiro, V.H.; Puga, H.; Meireles, J. Heat treatment as a route to tailor the yield-damping properties in A356 alloys. *Mater. Sci. Eng. A* **2018**, *729*, 1–8. [CrossRef]
28. Puga, H.; Barbosa, J.; Seabra, E.; Ribeiro, S.; Prokic, M. The influence of processing parameters on the ultrasonic degassing of molten AlSi9Cu3 aluminium alloy. *Mater. Lett.* **2009**, *63*, 806–808. [CrossRef]
29. Eskin, G.I. *Ultrasonic Treatment of Light Alloy Melts*; CRC Press: Boca Raton, FL, USA, 1998; ISBN 1-4987-0179-5.
30. Mozammil, S.; Karloopia, J.; Jha, P.K. Investigation of porosity in Al casting. *Mater. Today Proc.* **2018**, *5*, 17270–17276. [CrossRef]
31. Dispinar, D.; Akhtar, S.; Nordmark, A.; Di Sabatino, M.; Arnberg, L. Degassing, hydrogen and porosity phenomena in A356. *Mater. Sci. Eng. A* **2010**, *527*, 3719–3725. [CrossRef]
32. Puga, H.; Costa, S.; Barbosa, J.; Ribeiro, S.; Prokic, M. Influence of ultrasonic melt treatment on microstructure and mechanical properties of AlSi9Cu3 alloy. *J. Mater. Process. Technol.* **2011**, *211*, 1729–1735. [CrossRef]

metals

MDPI

Article

Optimizing the Conditions of Metal Solidification with Vibration

Olga Kudryashova [1,*], Marina Khmeleva [1], Pavel Danilov [1], Vladislav Dammer [1], Alexander Vorozhtsov [1] and Dmitry Eskin [1,2]

1 Faculty of Physics and Engineering, Tomsk State University, Tomsk 634050, Russia; khmelmg@gmail.com (M.K.); padanilov@gmail.com (P.D.); dammer.tomsk@yandex.ru (V.D.); abv1953@mail.ru (A.V.); dmitry.eskin@brunel.ac.uk (D.E.)

2 Brunel Centre for Advanced Solidification Technology, Brunel University London, Uxbridge UB8 3PH, UK

* Correspondence: olgakudr@inbox.ru; Tel.: +790-5924-5444

Received: 27 February 2019; Accepted: 20 March 2019; Published: 21 March 2019

Abstract: Vibration treatment of solidifying metals results in improvement in the ingot structure. There is a need to study this process not only because of the practical potential of vibration treatment but also due to the lack of understanding the process. An important practical challenge is to find optimal conditions for liquid metal processing. In this paper, the authors consider a solidification process in the particular case of a cylindrical chill mold with vibration as a solution of the Stefan problem. An integral value of mechanical stresses in the melt during solidification is considered as an efficiency criterion of vibration treatment. A dependence of this value on the vibration frequency and amplitude is obtained through solving the Stefan problem numerically. The solution allows one to find the optimal vibration frequency and amplitude. We verified the numerical solution with experimental data obtained upon vibration treatment of aluminum melt under different conditions. The experimentally found optimal conditions for metal processing were similar to those proposed in theory, i.e., a vibration frequency of about 60 Hz and an amplitude of about 0.5 mm.

Keywords: vibration treatment; optimal conditions; solidifying melt; Stefan problem

---

1. Introduction

Vibration treatment of a solidifying melt is one of the known methods of influencing the casting quality of aluminum alloys through reducing the grain size and obtaining a more uniform microstructure. An amplitude up to 5 cm and a frequency up to 10 kHz can be achieved with mechanical vibrations. At a higher frequency, e.g., above 16 kHz, ultrasonic treatment of melts can be performed [1]. Ultrasonic melt treatment of an A390 hypereutectic Al–Si alloy is shown to enhance the homogeneity of the microstructure and to increase the yield strength of the as-cast alloy by12% [2].

There are numerous theoretical and experimental works devoted to the influence of acoustic and mechanical vibrations on liquid metals during solidification (see for example [1] and [3–9]). It has been shown that vibration can modify the microstructure by multiplication of solidification substrates [5–7]. Study of the mechanical properties of A356 alloy after vibration treatment with frequencies of 100 Hz and 150 Hz showed the tensile and yield strengths improved by 20% and 10%, respectively, but the use of vibration frequencies of 200 Hz caused the formation of a high porosity microstructure and caused major defects [8]. The multiplication of grains is achieved through mechanical and thermal fragmentation of dendritic crystals by elastic stresses, microflows (thermal and solutal convection in a liquid) and cavitation [9–14]. Generally, the refinement of crystalline grains and the increasing soundness of a casting lead to improved mechanical properties and quality of cast products.

A mathematical model describing the vibration effect on liquid metals as a function of cavitation and turbulent flows in the bulk melt and in the solid-liquid zone has been proposed in [15].

Numerous prior studies showed the positive effect of both phenomena (cavitation and turbulence in the melt bulk) on the quality of casting microstructures [1,3,5]. These conditions can be implemented under high-intensity processes with relatively large amplitudes (more than 1 cm) and frequencies (more than 60 Hz but less than the ultrasonic range). In this case, cavitation in the liquid metal may occur in the melt saturated with gases (e.g., hydrogen in aluminum) during the solidification ("pseudocavitation") or under vibration treatment.

It was experimentally found [16] that the effect of grain refinement increases up to a certain frequency depending on the liquid alloy properties with an increase in the pre-ultrasonic vibration frequency. A further increase in the frequency reduces the effect as compared to an optimal vibration frequency (about 50 Hz with an amplitude of 0.49 mm in [16]). In our earlier work [15], the optimal frequency for high-intensity vibration treatment of melts was explained by the simultaneous occurrence of pseudocavitation and turbulent flows. The work [17], investigated the crystal behaviors under vibration using a transparent NH4Cl-70%H2O alloy (frequency from 20 to 1 000 Hz and acceleration from 10 to 100 m/s^2). The optimal frequency of 50 Hz and acceleration of 100 m/s^2 were found when the grain refinement effect was strengthened.

However, even in a more general case with low-intensity vibration, a small amount of gases in a melt, and a lack of cavitation and turbulence, vibration treatment of a solidifying alloy can also lead to a positive result [2,5]. In this case, there should also be an optimal frequency and amplitude under which the improvement of casting microstructure is most pronounced.

The aim of this study is to examine the mechanisms for improving the as-cast microstructures by vibration during alloy solidification; and based on a numerical solution of the Stefan problem to determine the optimal conditions (frequency and amplitude) for the vibration treatment without cavitation and turbulence.

2. Mathematical Model of Metal Solidification in a Cylindrical Volume with Vibration as a Stefan Problem

In considering solidification processes, an important issue is the local distribution of temperature field characteristics for a crystallizing ingot, such as temperature gradients and solidification-front velocity until the end of solidification.

The oscillating melt affects the stress distributions in the mushy (semi-solid) region of an ingot and influences the conditions of crystal growth. Under vibration, the moving melt rinses off saturated solute layers around growing crystals, increasing heat transfer and contributing to the dendrite growth. At the same time, the transfer of solute elements by a melt and their accumulation in interdendritic spaces can lead to local re-melting of the solid phase and the separation of dendritic branches, i.e., dendrite fragmentation [18]. The rate of heat flow with vibration treatment increases due to the convective heat transfer.

2.1. Model Setup

Let us consider the cooling and solidification of liquid metal in a thick-walled cylindrical chill mold subjected to low-frequency mechanical vibration (Figure 1). Mold walls are a source of horizontal vibration. Initially, the mold is uniformly heated to a temperature T_s, and a molten metal in the mold has a temperature T_0 above the liquidus temperature T_{m1}. Then the "liquid metal–hill mold" system cools down. The solidification starts at mold walls where the melt temperature at some point in time t_{col} becomes equal to T_{m1}. From this moment, the metal solidification starts and the solidification front moves from the mold walls to the cylinder axis. Solidification ends at a time t_{froz}. Then, the "solid metal–chill mold" system continues to cool down.

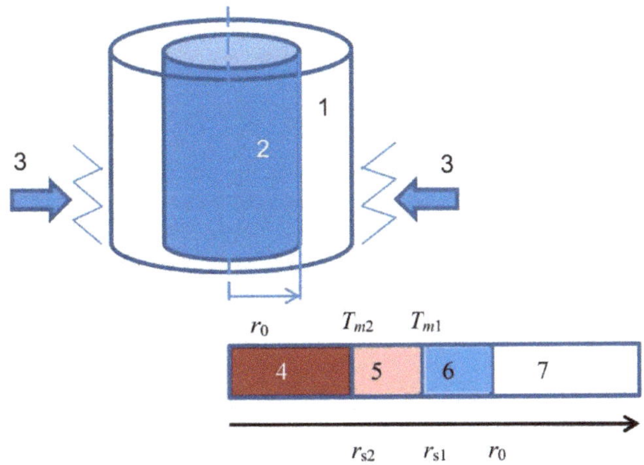

Figure 1. Scheme for the numerical setup of metal solidification with vibration: 1—thick-walled cylindrical chill mold, 2—liquid metal, 3—vibration generator, 4—molten zone, 5—transition zone, 6—zone of solid metal, 7—mold wall.

We made the following assumptions:

- The height of the chill mold is much larger than its inner diameter $2r_0$
- The heat transfer from the melt occurs only through the mold walls; the heat transfer at the top and bottom sides of the chill mold is ignored (an area of mold walls is much larger than the area of the mold bottom and open surface)
- The phase transition occurs within a temperature range from T_{m1} (liquidus temperature) to T_{m2} (solidus temperature)
- Microscopic volumes of the metal undergo longitudinal harmonic vibrations in an elastic wave with velocity v depending on the vibration frequency and amplitude.

The mathematical model of the phase transition under these assumptions is limited to solving a one-dimensional Stefan problem in cylindrical coordinates [19,20].

$$\begin{cases} \frac{\partial T_1}{\partial t} + v\frac{\partial T_1}{\partial r} = \frac{a_1}{r}\frac{\partial}{\partial r}\left(r\frac{\partial T_1}{\partial r}\right), & r \le r_{s2} \\ \frac{\partial T_{12}}{\partial t} = \frac{a_{12}(T)}{r}\frac{\partial}{\partial r}\left(r\frac{\partial T_{12}}{\partial r}\right), & r_{s1} > r > r_{s2} \\ \frac{\partial T_2}{\partial t} = \frac{a_2}{r}\frac{\partial}{\partial r}\left(r\frac{\partial T_2}{\partial r}\right), & r_0 > r \ge r_{s1} \\ \frac{\partial T_3}{\partial t} = \frac{a_3}{r}\frac{\partial}{\partial r}\left(r\frac{\partial T_3}{\partial r}\right), & r > r_0 \end{cases} \tag{1}$$

with initial and boundary conditions:

$$\begin{gathered} t = 0 : r_{s1} = r_0, T = T_0 \text{ under } r < r_0;\ T = T_s \text{ under } r = r_0 \\ r = r_{s2} : \frac{\partial T_1}{\partial r} = \frac{\partial T_{12}}{\partial r}, T_1 = T_{12} = T_{m2} \\ r = 0 : \frac{\partial T}{\partial r} = 0; r = r_0 : \frac{\partial T_2}{\partial r} = \frac{\partial T_3}{\partial r}, T_2 = T_3 \\ r \to \infty : T_3 = T_s \end{gathered} \tag{2}$$

where r is the cylindrical coordinate, t is the time, $a = \frac{c\rho}{\lambda}$ is the thermal diffusivity, index 1 refers to the melt, index 2 to the solid phase, index 12 to the transition zone, and index 3 refers to the chill mold.

The effective coefficient of thermal diffusivity in the transition zone is defined as $a_{12}(T) = a_1 + f_s a_2$, where f_s corresponds to the volume fraction of the solid phase in the transition (mushy) zone: $f_s = 1 - \frac{T - T_{m1}}{T_{m2} - T_{m1}}$.

The velocity of solidification front $\frac{\partial r_{s1}}{\partial t}$ is determined from the conditions at the border of solidification r_{s1} (that is, corresponding to the liquidus temperature T_{m1}) [19–21]:

$$r = r_{s1} : \lambda_1 \frac{\partial T_1}{\partial r} - \lambda_2 \frac{\partial T_2}{\partial r} = \rho_1 L \frac{dr_{s1}}{dt}, \; T_1 = T_{12} = T_{m1}, \tag{3}$$

where L is the latent heat of solidification, λ_1 and ϱ_1 are the thermal conductivity and liquid phase density, respectively, and λ_2 is the thermal conductivity of the solid phase.

Define an offset of fluid microscopic volumes in the environment by solving the set of wave equations:

$$\begin{cases} \frac{\partial^2 S_1}{\partial t^2} = \frac{c_1}{r} \frac{\partial}{\partial r}\left(r \frac{\partial S_1}{\partial r}\right), & r \le r_{s2} \\ \frac{\partial^2 S_{12}}{\partial t^2} = \frac{c_{12}(T)}{r} \frac{\partial}{\partial r}\left(r \frac{\partial S_{12}}{\partial r}\right), & r_{s1} > r > r_{s2} \\ \frac{\partial^2 S_2}{\partial t^2} = \frac{c_2}{r} \frac{\partial}{\partial r}\left(r \frac{\partial S_2}{\partial r}\right), & r \ge r_{s1} \end{cases} \tag{4}$$

with initial and boundary conditions:

$$\begin{array}{c} t = 0 : r_s = r_0, S = 0; \\ r = r_{s2} : \frac{\partial S_1}{\partial r} = \frac{\partial S_{12}}{\partial r}, S_1 = S_{12}; \\ r = r_{s1} : \frac{\partial S_{12}}{\partial r} = \frac{\partial S_2}{\partial r}, S_{12} = S_2; \\ r = 0 : \frac{\partial S_1}{\partial r} = 0; r = r_0 : S = A \sin \omega t, \end{array} \tag{5}$$

where S is the particle displacement relative to an equilibrium position, c is the sound speed, A and ω are the vibration amplitude and frequency, respectively. An effective speed of sound in the transition zone is defined as $c_{12}(T) = c_1 + f_s c_2$. The velocity of microscopic volumes of the liquid is calculated as: $v_1 = \frac{dS_1}{dt}$. The deformation and stress are defined as: $\varepsilon_1 = \frac{dS_1}{dr}$ and $\sigma_1 = E_1 \varepsilon_1 = E_1 \frac{dS_1}{dr}$, respectively, where E_1 is the volumetric modulus of melt elasticity which is determined from the known ratio in the longitudinal wave: $E_1 = c_1^2 \rho_1$. The values v_{12}, ε_{12}, σ_{12} are also calculated in the transition zone.

Consider an integral characteristic of melt stresses during the time of solidification:

$$Z_\sigma = \frac{1}{r_0} \int_{t_{col}}^{t_{froz}} |\sigma_1|. \tag{6}$$

Z_σ has a dimensionality of mechanical impedance $Z_s = c_1 \varrho_1$. This value can be considered as a stress integral during solidification related to the unit element of solidification front trajectory from the mold walls to its axis r_0. It also can be considered as the total value of stresses of the vibration field directed at overcoming the mechanical impedance during solidification.

It can be assumed that the integral value Z_σ determines the effectiveness of vibration and provides a comparative characteristic for evaluating process conditions.

2.2. Modeling Results

Equations from (1) to (5) were solved numerically using an explicit difference scheme and a fixed grid method [22] implemented in Delphi 7.0.

The physical properties for an A356 alloy (Al-7% Si), listed in Table 1, were used in solving the problem.

Table 1. Physical quantities used in calculation.

Properties	Density ϱ, kg/m^3 [23,24]	Specific Heat c, J/(kg·°C) [24,25]	Thermal Conductivity λ, W/(m·°C) [24,25]	Elastic Modulus E 10^{-5}, MPa [23]	Sound Speed c_s, m/s [23]	Liquidus Temperature T_{m1}, K [26]	Solidus Temperature T_{m2}, K [26]
Melt (1)	2362	1177	98.1	0.52	4700	883	841
Solid metal (2)	2660	880	155.0	0.70	6260	-	-
Steel (3)	7800	462	50.2	-	-	-	-

The ranges of process parameters in the calculation were as follows:

- Frequency f = 0–100 Hz
- Amplitude A = 0.1–10 mm
- Initial temperature of the chill mold T_s = 430–630 K
- Initial temperature of the liquid metal T_0 = 900–1050 K
- Radius of chill mold was 17.5 mm
- Specific heat of phase transition (latent heat of solidification) was L = 429 kJ/kg [24].

The calculation showed that the temperature profile weakly depends on vibration within the given range of amplitude and frequency. Figure 2a shows the temperature in its center $T(0,t)$ without and with vibration at a frequency of 50 and 80 Hz. The movement of the solidification front r_{s1} (that is, corresponding to the liquidus temperature T_{m1}) is nonlinear (Figure 2b).

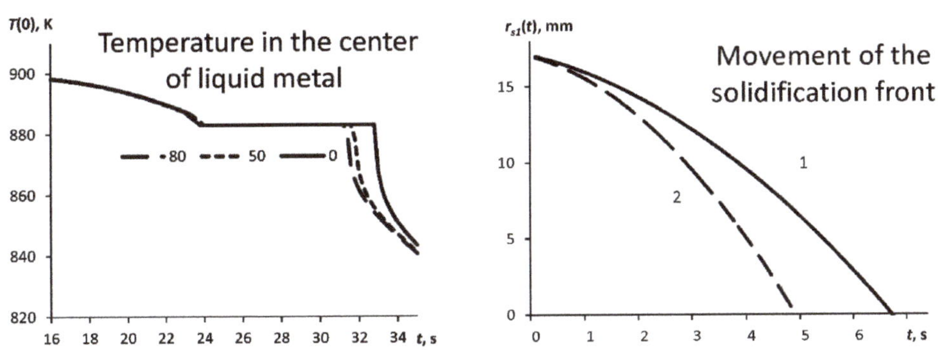

(a) (b)

Figure 2. (**a**) Time dependence of the temperature in the center of liquid metal volume for different conditions: 0—without vibration, 50—vibration of f = 50 Hz, 80–vibration of f = 80 Hz; T_s = 630 K, T_0 = 900 K.; (**b**) Time dependence of the movement of the solidification front from the start point t_{col} for two thermal modes: 1—T_s = 630 K, T_0 = 900 K, and 2—T_s = 430 K, T_0 = 1050 K under A = 0.5 mm, f = 50 Hz.

First of all, the thermal solution allows one to define the kinetics of the process. Figure 2a shows the time-dependence of the temperature in the center of the melt volume $T(0,t)$. Vibration accelerates the process of solidification: the "plateau" on the temperature profile corresponding to the solidification becomes shorter with the vibration.

The calculation shows that the duration of alloy solidification as well as the cooling time (combined cooling to the temperature of liquidus and then to the solidus), depends on the initial temperature difference between the melt and the chill mold, and is less dependent on frequency (Figure 3a). The solidification time slightly shortens with the increasing frequency (Figure 3b). As one can see, the solidification time is short in all cases, a few seconds. However, it is during this brief period of time that the vibration has an effect on the solidification and structure formation.

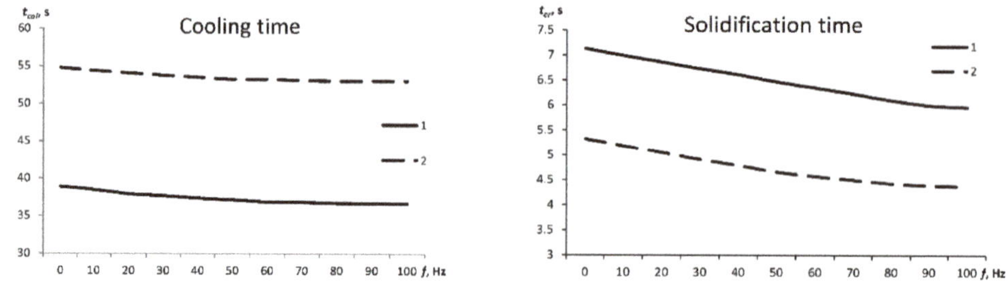

(a) (b)

Figure 3. Dependencies of the cooling time (**a**) and the solidification time (**b**) on the vibration frequency: 1—T_s = 630 K, T_0 = 900 K, and 2—T_s = 430 K, T_0 = 1050 K; A = 0.5 mm.

Figure 4 shows the dependence of the cooling time t_{col} on the initial mold temperature, as well as the dependence of the solidification time on the initial melt temperature. The higher the initial temperature of the metal and that of the chill mold, the more time needed for cooling; which is the expected result.

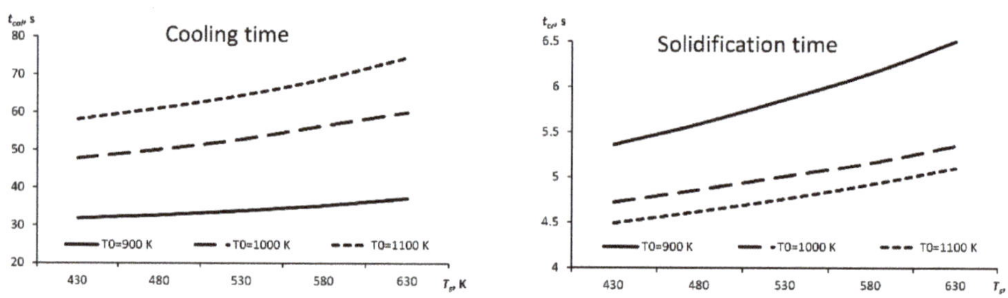

(a) (b)

Figure 4. Dependence of (**a**) the cooling time before solidification and (**b**) the solidification time on the initial temperature of the chill mold T_s for different starting melt temperatures T_0.

When analyzing the stress amplitude imposed on the semi-solid alloys, one should note that the strength of semi-solid alloys can be significantly lower than that of the solid alloy, i.e., the tensile strength of an A356 alloy decreases from 157 MPa in the solid state [27] to 4.98 MPa at 860 K and to ~ 0.01 MPa at 880 K [24]. Nevertheless, a certain critical amplitude of mechanical stress σ_{min} that is higher than the tensile strength of the semi-solid alloy, is required to effect the fracture of growing crystals, and as a consequence, the refinement of the crystalline grains in the casting. Not only the value of σ_{min} is important but also the operating time of these stresses during the alloy solidification (number of oscillations during solidification). A combination of those can be characterized by the value Z_σ (Equation (6)).

Figure 5a shows the dependence of the specific stress integral Z_σ on the vibration frequency for the two following conditions: T_s = 630 K, T_0 = 900 K and T_s = 430 K, T_0 = 1050 K. The calculation demonstrates that the effectiveness of the vibration grows nonlinearly with increasing the vibration frequency and saturates at a certain frequency. This effect can be explained by the reduction of the solidification time during which the vibration affects the growing crystals (see Figure 3b). It can be assumed that the fracture of growing crystals is possible only under conditions when a relative value Z exceeds a certain critical value: $Z = Z_\sigma/Z_s = Z_\sigma/c_1\rho_1 > Z_{cr}$.

The dependence of this integral stress characteristic on amplitude is, however, linear and the values become substantial at the frequencies above 50 Hz (Figure 5b).

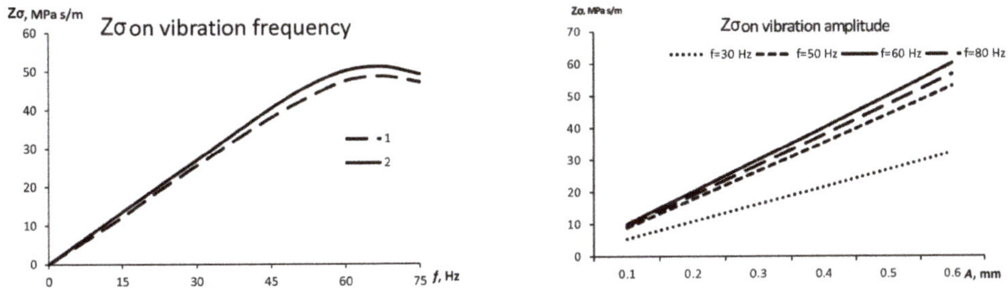

(a) (b)

Figure 5. (**a**) The dependence of the specific stress integral on the vibration frequency for two thermal conditions: 1—T_s = 630 K, T_0 = 900 K, and 2—T_s = 430 K, T_0 = 1050 K; A = 0.5 mm. (**b**) The dependence of the specific stress integral on the vibration amplitude for the four values of frequency.

Thus, according to the proposed approach, the vibration of liquid and solidifying metal causes the following effects:

1. Acceleration of cooling and solidification as a result of convective heat transfer regardless of little change in the temperature profile.
2. The generation of mechanical stresses in the solid–liquid zone that can potentially fracture (fragment) the growing crystals.

Considering that the vibration stresses operate on growing crystals in a particular short period of solidification, the time integral characteristic of mechanical stresses can be suggested as a measure of the effectiveness of vibration. The higher this value, the greater the effect of vibration stress on crystals. The integral characteristic of mechanical stresses increases linearly with rising amplitude but grows nonlinearly with rising vibration frequency. It is necessary to consider the time and mechanical conditions under vibration treatment of an alloy during solidification as:

- Vibration affects the structure from the beginning of solidification and until its end.
- The higher the vibration amplitude, the higher the vibration effect (it grows linearly).
- There is the optimal vibration frequency at which this effect is the highest. This frequency equals 60 Hz for parameters used in the calculation (Table 1).

3. Experimental Verification of the Mathematical Model by Casting of an Aluminum Alloy with Vibration

3.1. Experimental Procedures

A vibrating table EV 341-07 (PC "Lighthouse YF", Yaroslavl, Russia) was used to carry out the experiment to evaluate the vibration processing of solidifying melt. The scheme of the experimental setup is shown in Figure 6. Studies were performed on an A356 aluminum alloy.

The experimental technique for vibration treatment of the aluminum melt was as follows. The A356 aluminum alloy was placed in a crucible inside an electrical melting furnace (Nabertherm GmbH, Lilienthal, Germany) at a furnace temperature of 1073 K. A steel chill mold (with an internal diameter of 35 mm, wall thickness of 135 mm, height of 200 mm), preheated to a given temperature 430 K, was fixed on a vibrating table that oscillated in a mode of horizontal vibration (the same condition as used in the model described above). Then, using a holding device we removed the crucible from the furnace and poured the liquid alloy in the preheated chill mold at a temperature of 973 K (700 °C). The fixed vibration frequencies were 50, 60 and 80 Hz in order to cover the range around the maximum in Figure 5a. The vibration amplitude was varied in the range from 0.38 to 0.53 mm (angle of an off-center block is from 10° to 20°, the link between amplitude and the angle θ [rad] is expressed as follows: $A = \frac{\pi-\theta}{\pi} A_{\max}$, where A_{max} = 0.56 mm). The vibration continued for

1–2 min after pouring until the complete solidification of the melt. A reference experiment without vibration (with the other parameters the same) was conducted to assess the effectiveness of vibration (grain structure and density were chosen as the metallurgical indicators).

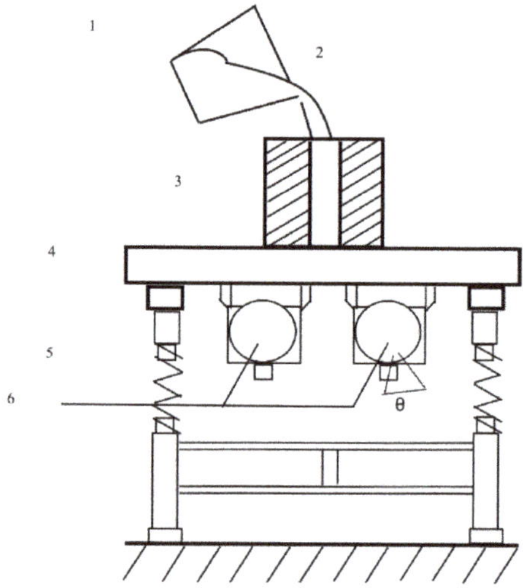

Figure 6. Scheme of the experimental setup: 1—crucible; 2—liquid metal; 3–chill mold; 4—table; 5—springs; 6—vibrators.

3.2. Analytical Techniques

The structure of the obtained materials was investigated using optical microscopy (Olympus GX-71, Tokyo, Japan) and scanning electron microscopy (Philips SEM-515, Hillsboro, Oregon, USA). A universal testing machine Instron 3369 was used for mechanical tensile testing of the obtained alloys. Three samples were tested for each condition. The density of the obtained alloys was estimated by a hydrostatic weighing method (using Archimedes' principle).

Samples were pre-polished on a circular grinding machine (Buehler, Lake Bluff, IL, USA) and then subjected to mechanical polishing. Samples were etched using a 5% HBF4 acid solution to identify the microstructure. Mechanical tensile tests are carried out with a tensile test machine Istron 3369 (Norwood, MA, USA). The samples were cut in the form of plates (length of the working part 25 mm, width 6 mm, thickness 2 mm, rounded radius 18 mm) using electroerosion cutting of aluminum alloy. Tensile testing are conducted at a rate of 0.2 mm/min until a fracture occurs. Test results were obtained in the form of "stress-strain" diagrams from which the values of yield strength, tensile strength and maximum elongation were found.

4. Results and Discussion

At a frequency of 60 Hz and the angle of off-center block θ = 20° (amplitude of 0.53 mm), the density of samples cut from the lower part of an ingot increases in comparison with the density obtained at 50 Hz (Table 2).

Increasing the density of samples may be linked to the facilitated gas (hydrogen) release during melt vibration that decreases gas and shrinkage porosity [5,28]. A further increase in the vibration frequency up to 80 Hz leads to a decrease in the density of an ingot. A similar trend is observed at the angle of off-center block θ = 10° (amplitude of 0.38 mm), i.e., the increased frequency 80 Hz slightly reduces the density of the samples.

These measurements correlate well with the modeling (Figure 5a): the integral of mechanical stresses increases with the vibration frequency up to 60 Hz and then slightly decreases to 80 Hz. This is due to the reduction in the solidification time with increasing frequency as a result of intensified thermal processes in the melt, which consequently reduces the effective time of vibration and for degassing.

The observation of the macrostructure of the obtained A356 ingots without (Figure 7a) and with vibration (Figure 7b) shows that vibration processing (frequency of 60 Hz and amplitude of 0.5 mm) leads to a significant grain refinement.

Table 2. Measurements of the density of A356 aluminum alloy.

No. of Sample	Frequency, Hz	Amplitude, mm (θ, °)	Density, g/cm^3
1 2 3 4	50 Hz	0.53 mm (20 °)	2.70 ± 0.02
1 2 3 4	60 Hz		2.72 ± 0.03
1 2 3 4	80 Hz		2.69 ± 0.03
1 2 3 4	50 Hz	0.38 mm (10 °)	2.69 ± 0.04
1 2 3 4	60 Hz		2.70 ± 0.02
1 2 3 4	80 Hz		2.69 ± 0.02

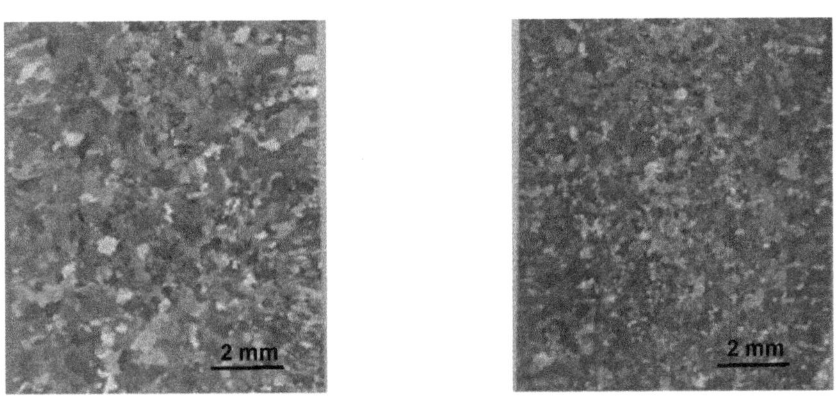

(a) (b)

Figure 7. Macrostructure of an A356 ingot: (**a**) obtained without vibration; (**b**) obtained with vibration during solidification (frequency of 60 Hz and amplitude of 0.5 mm).

The microstructures of the A356 alloy samples (Figure 8) show that significant structural changes occurred during the vibration. The average grain size reduces from 208 μm in the alloy produced without vibration to 89 μm for the alloy cast with vibration (Figure 9).

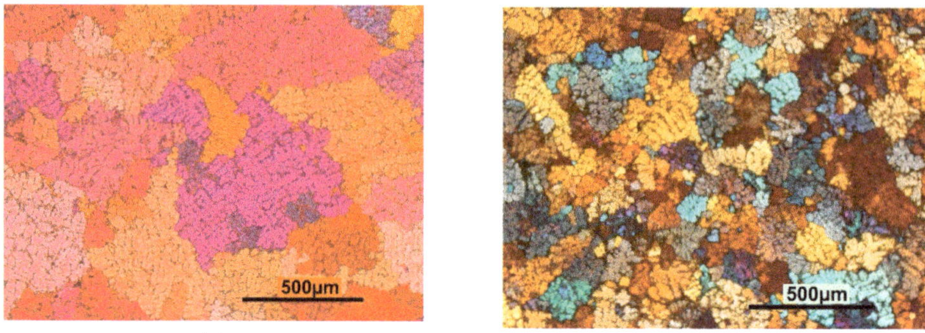

(a) (b)

Figure 8. Microstructure of an A356 aluminum alloy: (**a**) without vibration; (**b**) after vibration during solidification (frequency of 60 Hz and amplitude of 0.5 mm).

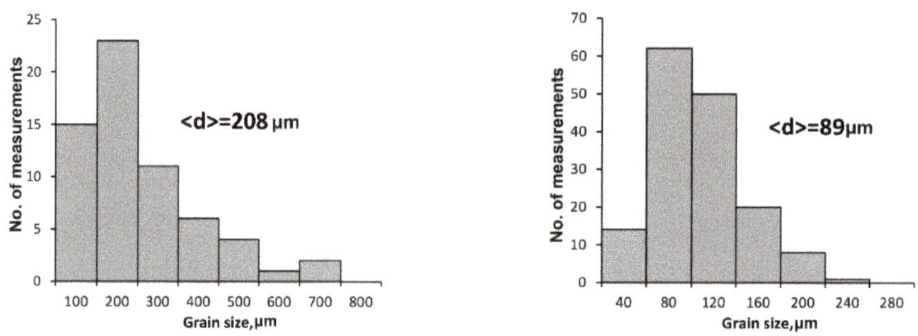

(a) (b)

Figure 9. Grain size distribution in an A356 aluminum alloy (**a**) without vibration and (**b**) after vibration during solidification (frequency of 60 Hz and amplitude of 0.5 mm).

Based on the obtained experimental and theoretical results, we can conclude that the optimal frequency for vibration processing for the given mold geometry and the range of parameters is 60 Hz. The calculations also showed that the higher the amplitude, the better the result of the treatment. The experimental study showed a similar trend with the maximum density of the samples obtained at the highest amplitude used (0.53 mm, angle of 20°).

The improvement of the structure in the castings obtained with vibration is related to intensive heat transfer in the melt. The vibration energy is spent on the fragmentation of dendritic branches, and this process results in grain multiplication. The alloy structure obtained by vibration shows the presence of very small grains (Figure 8b) which represent preserved dendrite fragments.

Mechanical testing of the as-cast A356 aluminum alloy shows that yield strength $\sigma_{0.2}$ increases from 67 ± 6 to 121 ± 7 MPa after vibration of the melt with a frequency of 60 Hz and an amplitude of 0.53 mm compared to the sample without vibration. This result can be associated with the decrease in the grain size and porosity and the increased density of the samples. Meanwhile, the tensile strength remains unchanged at the level $\sigma_B = 182 \pm 7$ MPa and the elongation δ remains at the level $\delta = 3.4 \pm 0.2\%$ (Table 3, Figure 10).

Table 3. Mechanical testing of A356 aluminum alloy.

Frequency, Hz	σ_B, MPa	δ, %
Without vibration	150 ± 7	2.1 ± 0.1
50	160 ± 6	2.3 ± 0.1
60	182 ± 7	3.4 ± 0.2
80	149 ± 6	2.1 ± 0.1

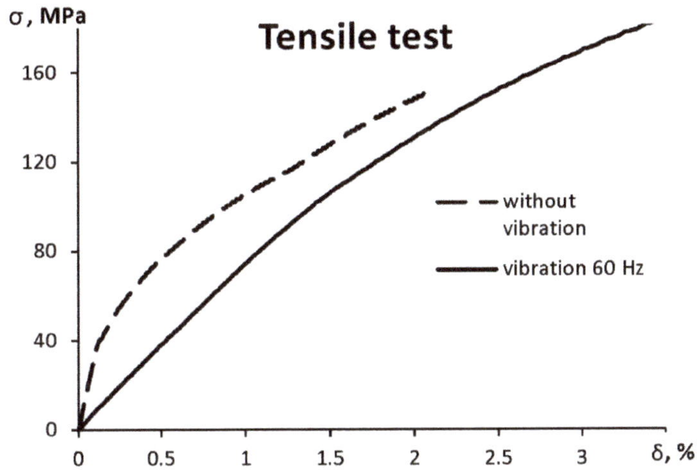

Figure 10. Stress-strain diagrams of the A356 aluminum alloy without vibration and after vibration during solidification (frequency of 60 Hz and amplitude of 0.5 mm).

Our results agree well with the work of Murakami et al. [17] who studied the effect of vibration frequency, acceleration amplitude, velocity amplitude, and displacement amplitude on the size and shape of the grains in an JIS AC4CH aluminum alloy (A356 analog). They used the frequencies of 10, 20, 50,100,150 and 200 Hz and found that the smallest grains (132 μm) were obtained at a frequency of 50 Hz and an amplitude of 0.49 mm, which is close to our result. However, their work did not cover the frequency range of our interest, i.e., 50 to 80 Hz.

It is important to note that in the frequency range 50–80 Hz and amplitudes up to 1 mm there are no conditions for cavitation and turbulent fluid flow [15]. At the same time, there are no conditions for entrapment of atmospheric gases and oxide films, which leads to extremely undesirable porosity in the ingot. When implementing more intensive and high-frequency process conditions, both the positive effects of turbulence and cavitation (better mixing, higher stresses, efficient degassing) and the negative ones (entrapment of gases and oxides leading to pores and cavities in the metal) should be considered.

5. Conclusions

In this study we showed that the experimentally observed improving of an as-cast structure at low-frequency and low-intensity vibration can be reasonably described by the thermal model (Stefan problem with mechanical vibration).

Optimal conditions for low-frequency and low-intensity vibration of the melt in a cylindrical chill mold were determined by theoretical calculation and verified by experimental testing, that is, a vibration frequency of about 60 Hz and an amplitude of about 0.5 mm. Under such conditions, there is a significant decrease in the grain size in the ingot as well as improved soundness (density) of the as-cast metal. The yield strength increased by ~80% as compared to the ingot cast without vibration.

The optimal conditions within the thermal model are explained by the reduction of the solidification time with increasing frequency on the one hand, and by the growing integral mechanical stresses during solidification on the other hand. The specific integral of mechanical stresses in the melt subjected to vibration during solidification can be considered as a measure of the effectiveness of vibration within the thermal approach for the process description.

Theoretical calculation allows one to optimize the time of the vibration effect. The vibration has no practical value if applied above the liquidus temperature. Moreover, it can act as a source of the undesirable phenomena of gas capture. The vibration should be started just at the beginning of solidification and finish with the complete metal solidification.

The mathematical model presents a cylindrical mold with horizontal vibration. The model does not take into account other geometric variations, nor the vertical component of vibration. In future, it would be useful to simulate other conditions of metal solidification with vibration.

Author Contributions: Conceptualization, A.V. and D.E.; Data curation, V.D. and D.E.; Formal analysis, O.K.; Funding acquisition, A.V.; Investigation, D.E.; Methodology, O.K. and P.D.; Project administration, A.V.; Resources, V.D.; Software, O.K.; Supervision, A.V.; Validation, M.K. and V.D.; Writing—original draft, O.K.; Writing—review & editing, M.K. and P.D.

Funding: This research was funded by Russian Science Foundation, grant number 17-13-01252.

Conflicts of Interest: The authors declare no conflict of interest.

References

1. Eskin, G.I.; Eskin, D.G. *Ultrasonic Treatment of Light Alloy Melts*, 2nd ed.; CRC Press: London, UK; New York, NY, USA, 2014.
2. Kim, S.B.; Cho, Y.H.; Jung, J.G.; Yoon, W.H.; Lee, Y.K.; Lee, J.M. Microstructure-Strengthening Interrelationship of an Ultrasonically Treated Hypereutectic Al–Si. *Met. Mater. Int.* **2018**, *24*, 1376–1385. [CrossRef]
3. Balandin, G.F. *Formation of the Crystalline Structure of Castings: Pure Metals and Single-Phase Alloys*; Mashinostroyeniye: Moscow, Russia, 1973. (In Russian)
4. Shukla, D.P.; Goel, D.B.; Pandey, P.C. Effect of vibration on the formation of porosity in aluminum alloy ingots. *Metall. Trans. B* **1980**, *11*, 166–168. [CrossRef]
5. Campbell, J. Effects of vibration during solidification. *Int. Met. Rev.* **1981**, *26*, 71–108. [CrossRef]
6. Fisher, T.P. Effects of vibrational energy on the solidification of aluminum alloys. *Br. Foundrym.* **1973**, *66*, 71–84.
7. Burbure, R.R.; Hareesha, I.; Murthy, K.S.S. Influence of Low-Frequency Vibrations on Aluminum Eutectics. *Br. Foundrym.* **1979**, *72*, 34–38.
8. Selivorstov, V.; Dotsenko, Y.; Borodianskiy, K. Influence of low-frequency vibration and modification on solidification and mechanical properties of Al-Si casting alloy. *Materials* **2017**, *10*, 560. [CrossRef]
9. Jiang, W.; Chen, X.; Wang, B.; Fan, Z.; Wu, H. Effects of vibration frequency on microstructure, mechanical properties, and fracture behavior of A356 aluminum alloy obtained by expendable pattern shell casting. *Int. J. Adv. Manuf. Technol.* **2016**, *83*, 167–175. [CrossRef]
10. Taghavi, F.; Saghafian, H.; Kharrazi, Y.H.K. Study on the effect of prolonged mechanical vibration on the grain refinement and density of A356 aluminum alloy. *Mater. Des.* **2009**, *30*, 1604–1611. [CrossRef]
11. Chirita, G.; Stefanescu, I.; Soares, D.; Silva, F.S. Influence of vibration on the solidification behavior and tensile properties of an Al–18 wt% Si alloy. *Mater. Des.* **2009**, *30*, 1575–1580. [CrossRef]
12. Zheng, Z.; Ji, Y.J.; Mao, W.M.; Rui, Y.U.E.; Liu, Z.Y. Influence of rheo-diecasting processing parameters on microstructure and mechanical properties of hypereutectic Al–30% Si alloy. *Trans. Nonferrous Met. Soc. China* **2017**, *27*, 1264–1272. [CrossRef]
13. Olufemo, A.F.; Asemola, I.S. Effects of melt vibration during solidification on the mechanical property of Mg-Al-Zn alloy. *Int. J. Metall. Eng.* **2012**, *1*, 40–43.
14. Wang, W.L.; Wang, K.S.; Lin, X. Real time observation of nucleation and detachment of crystal from chilling surface Part I–effects of vibration. *Int. J. Cast Met. Res.* **2010**, *23*, 344–348. [CrossRef]
15. Vorozhtsov, S.; Kudryashova, O.; Promakhov, V.; Dammer, V.; Vorozhtsov, A. Theoretical and experimental investigations of the process of vibration treatment of liquid metals containing nanoparticles. *JOM* **2016**, *68*, 3094–3100. [CrossRef]

16. Murakami, Y.; Miwa, K.; Kito, M.; Honda, T.; Kanetake, N.; Tada, S. Effects of Mechanical Vibration Factors on Size and Shape of Solid Particles in JIS AC4CH Aluminum Alloy Semi-Solid Slurry. *Mater. Trans.* **2016**, *57*, 163–167. [CrossRef]
17. Ruvalcada, D.; Mathiensen, R.H.; Eskin, D.G.; Arnberg, L.; Katgerman, L. In situ observations of dendritic fragmentation due to local solute-enrichment during directional solidification of an aluminum alloy. *Acta Mater.* **2007**, *55*, 4287–4292. [CrossRef]
18. Zhang, Y.; Meng, W.A.N.G.; Du, L.C.; Xin, L.I.N.; Huang, W.D. In-situ observation of crystal detachment behavior in NH4Cl-H2O system from chilling metal surface with vibration. *Trans. Nonferrous Met. Soc. China* **2010**, *20*, s821–s825. [CrossRef]
19. Meirmanov, A.M. *The Stefan Problem*; Walter de Gruyter: Berlin, Germany, 1992; Volume 3.
20. Kamenomostskaya, S.L. On Stefan's problem. *Mat. Sb.* **1961**, *53*, 489–514. (In Russian)
21. Alexiades, V.; Cannon, J.R. Free boundary problems in solidification of alloys. *SIAM J. Math. Anal.* **1980**, *11*, 254–264. [CrossRef]
22. Hu, H.; Argyropoulos, S.A. Mathematical modelling of solidification and melting: A review. *Model. Simul. Mater. Sci. Eng.* **1996**, *4*, 371. [CrossRef]
23. Zinoviev, V.E. *Thermophysical Properties of Metals at High Temperatures. Reference Book*; Metallurgiya: Moscow, Russia, 1989. (In Russian)
24. Novikov, I.I. *Hot Shortness of Non-Ferrous Metals and Alloys*; Nauka: Moscow, Russia, 1966. (In Russian)
25. Bakhtiyarov, S.I.; Overfelt, R.A.; Teodorescu, S.G. Electrical and thermal conductivity of A319 and A356 aluminum alloys. *J. Mater. Sci.* **2001**, *36*, 4643–4648. [CrossRef]
26. Babichev, A.P.; Babushkina, N.A.; Bratkovskiy, A.M. *Physical Quantities: Reference Book*; Energoatomizdat: Moscow, Russia, 1991. (In Russian)
27. Arnberg, L.; Backerud, L.; Chai, G. *Solidification Characteristics of Aluminum Alloys: Dendrite Coherency*; American Foundrymen's Society: Des Plaines, IL, USA, 1996; Volume 3.
28. Kocatepe, K. Effect of low frequency vibration on porosity of LM25 and LM6 alloys. *Mater. Des.* **2007**, *28*, 1767–1775. [CrossRef]

metals

MDPI

Article

Joint Effect of Ultrasonic Vibrations and Solid Metal Addition on the Grain Refinement of an Aluminium Alloy

Dmitry Eskin [1,2,*] **and Feng Wang** [1]

[1] BCAST, Brunel University London, Uxbridge UB8 3PH, UK; feng.wang@brunel.ac.uk
[2] Faculty of Physics and Engineering, Tomsk State University, Tomsk 634050, Russia
* Correspondence: dmitry.eskin@brunel.ac.uk; Tel.: +44-1895-265317

Received: 21 December 2018; Accepted: 27 January 2019; Published: 1 February 2019

Abstract: It is well known that it takes some time for the solid phase to completely dissolve upon melting, especially inside the defects of insoluble particles, e.g. oxides. Until then the oxides remain active solidification substrates in the case of subsequent solidification. It is also known that ultrasonic melt treatment causes grain refinement through activation and dispersion of solidification substrates (one of the mechanisms) and also accelerates the dissolution of solid metal in the melt. In this study we combine these effects and demonstrate that the introduction of an alloy rod into the matrix melt of the same composition results in significant grain refinement, this effect is increased by the ultrasonic vibration of the rod. The achieved grain size is comparable to that obtained by a standard Al–Ti–B grain refiner. All samples were cast using a standard TP-1 mould to enable correct comparison. The effects of the temperature range of the rod introduction as well as the application of ultrasonic vibrations are discussed.

Keywords: aluminium alloy; grain refinement; ultrasonic processing; solidification; nucleation; dissolution

1. Introduction

The fact that by introducing solid metal into the melt one can achieve structure refinement is rather well known. As early as in the 1930s–1950s a series of papers and patents have been published, showing the benefits for grain refinement when dissolving solid metal in the melt before solidification [1,2]. Later on, this way of structure refinement was developed further and dubbed "suspension casting" or introduction of "internal chills" [3,4]. The solid metal was added in a form of cut wire, cut sheet, or powder, with the resulting grain refining, elimination of columnar grain structure in ingots and castings of steel, copper and aluminium alloys. The underlying mechanisms have been suggested as (1) rapid cooling of the melt due to the latent heat consumption upon melting of the solid metal with the resultant melt undercooling and (2) introduction of many solidification substrates in the form of crystal fragments and active non-metallic inclusions [1,5]. A number of patents have been filed where either the same solid alloy or a master alloy with additions (e.g. Ti for Al alloy) is introduced in the amounts up to 50% (typically less than 10%) into the melt close to the liquidus temperature [6–9]. The main reason why this technique is not widely used in industry is the possible incomplete dissolution of the solid parts introduced into the melt with ensuing inhomogeneous as-cast structure and potential defects. In addition, the selection of the temperature range where the technology works the best is not clear.

Another well-documented means of controlling the grain structure of as-cast metals is ultrasonic melt processing (USP) [10–13]. There are well-studied mechanisms through which USP affects the structure, i.e., wetting of non-metallic inclusions, cavitation-assisted heterogeneous nucleation, fragmentation of primary intermetallics and dendrites and enhanced mixing of the liquid volume [13]. Although very powerful in grain refinement, this technology suffers from instrumental issues, i.e., the ultrasonic tool (sonotrode) that is used for direct introduction of high-frequency vibration into the melt is subject to cavitation erosion and (if made of metal) gradual dissolution. The choice of the sonotrode material is, therefore, very important and it has been demonstrated that Nb-based alloys are most stable under cavitation condition in the aluminium melt [10]. But even sonotrodes made from Nb alloys suffer eventually from erosion. At the same time, some schemes that use a "consumable" sonotrode have been suggested in ultrasonic welding (wire feeding) and in electro-slag remelting (vibrating electrode) as reviewed elsewhere [11,12]. Such a scheme opens some new avenues in ultrasonic melt processing as it eliminates the issue of sonotrode material selection. It is also well known that ultrasonic vibrations significantly accelerate the dissolution of solid metals in the melt as well as promote rapid mixing of solutes through accelerated diffusion, eliminating the build-up of the solute-rich layer at and facilitating a better excess of the fresh melt to the solid/liquid phase [10,14].

In this study, we attempted to combine these two technologies to achieve a synergetic effect.

2. Experimental Methods and Materials

An AA6082 alloy was used as a base material (1.16% Si, 0.7% Mg, 0.98% Mn, 0.26% Fe, 0.17% Cr, 0.04% Ti) and a 9-mm rod of the same alloy was used for additions. In some experiments, a standard Al5Ti1B rod (LSM, Rotherham, UK) was used (0.2% addition). A total of 10% of AA6082 rod was introduced in each experiment. The choice of the alloy is due to its wide applicability in automotive applications where extrusions are made from direct-chill case billets, so the grain refinement is an important issue. The liquidus of this alloy is 647 °C, the equilibrium solidus is 593 °C and the nonequilibrium solidus is 533 °C, as calculated by Thermocalc software (version 2019a, TCAL 4 database, Thermo-Calc Software AB, Solna, Sweden).

In the experiments with ultrasound, a 5-kW ultrasonic generator and a 5-kW, 17.5-kHz water-cooled transducer (Reltec, Yekaterinburg, Russia) were used at a power of 3 kW (the longitudinal amplitude was ±12 μm). When ultrasound was applied, the rod was inserted tightly into a hole in a steel sonotrode in the half-wavelength position and then fed into the melt while the ultrasonic transducer was on as shown in Figure 1. The sonotrode was not in contact with the melt and was not cooled or heated. As a result of this scheme, the longitudinal oscillations of the sonotrode were transformed into flexural oscillations of the rod with the resultant shorter wavelength and larger amplitudes, up to 25 μm as measured by a contactless vibrometer. Separate experiments in water showed that cavitation conditions were achieved at the tip of the rod as well as in nodes along the rod length. Water is a well-accepted analogue of liquid aluminium with regard to the cavitation behaviour, though the cavitation threshold in liquid Al is about two times higher than in water, 10 μm vs 5 μm peak-to-peak amplitude at 20 kHz [15]. Therefore, the vibration conditions used in these experiments were sufficient to achieve cavitation in liquid Al.

The melt was prepared in clay-graphite crucibles (0.5 kg charge) in an electric resistance furnace (Carbolite, Hope Valley, UK) with melt temperature up to 770 °C. Rod introduction was made either without or with ultrasonic vibrations applied into the melt in a temperature range with the final pouring temperature 670 °C, i.e., the rod was started to be introduced at a certain given melt temperature and then, while the rod was dissolving, the melt was naturally cooled down to the pouring temperature. Reference samples were prepared by adding the same amount of rod into the melt that was then superheated to 750 °C, left in the furnace for 30 min and then cooled in air and cast at 670 °C. The rod was not preheated before the introduction. Casting was done in a standard TP-1 mould [16].

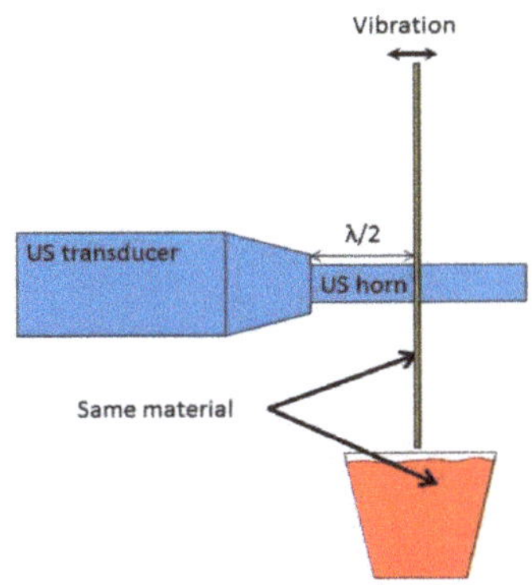

Figure 1. A schematic of the experiment.

The castings were then sectioned along the vertical centre plane and etched for structure examination. Macroetching was done in a 10% NaOH water solution after grinding. To reveal grain microstructure smaller samples were cut, ground and polished using standard procedures and then anodised in a 5% HFB_4 water solution at 20 VDC. The grain structure was subsequently examined in an optical microscope Zeiss Axioscope (Zeiss, Cambridge, UK) in polarised light. For the grain size measurement, the sections in the middle of the sample cross-section were used (see Figure 1a for detail). The grain size was measured using a random linear intercept method with at least 50 grains measured. Statistical analysis of the measured grain size was performed and the average values are reported.

3. Results and Discussion

Reference samples (a number of those had been produced) showed a coarse columnar grain structure as shown in Figure 2 with the average grain size 630 μm.

First experiments when the rod was introduced without or with ultrasound in the temperature ranges 740 to 715 °C and 740 to 710 °C did not produce any significant results with a very coarse mixed columnar and equiaxed structure formed (these results are not presented here). This confirmed the importance of the temperature range needed for the desired effect. Although the introduced rod was quick to dissolve, the relatively high temperature apparently prevented the survival of solid fragments as well as facilitated deactivation of solid inclusions (only 10–15 °C above liquidus is needed for that [9]).

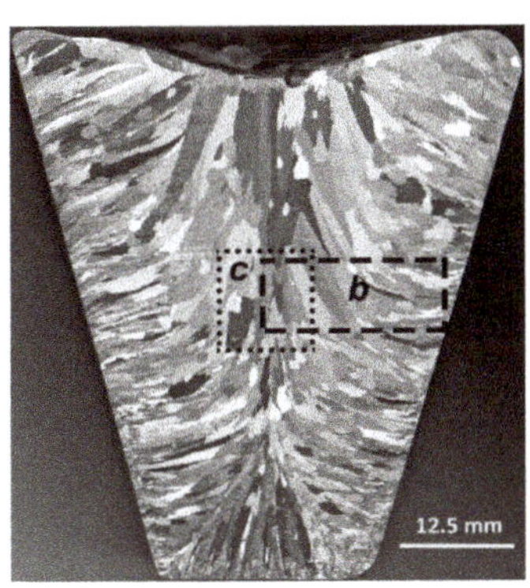

(a)

Figure 2. *Cont.*

(b)

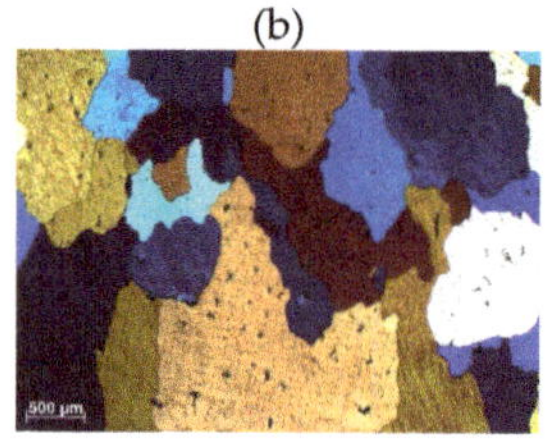

(c)

Figure 2. Macro (**a**) and micro (**b**,**c**) grain structure of a reference sample cast at 670 °C in a TP1 mould. The positions of photos in (**b**) and (**c**) with respect to the sample cross-section are marked in (**a**) (same positions have been used hereafter).

Following experiments were performed in the temperature ranges 720 to 670 °C and 710 to 670 °C. The decrease in the introduction temperature yielded a remarkable change in the grain structure even when the rod was introduced without ultrasonic vibrations, with the average grain size 250 µm, Figure 3. It can be noted that a slight decrease in the starting melt temperature gives some additional grain refining effect (228 µm), most probably due to the better survival chances of crystal fragments and active inclusions.

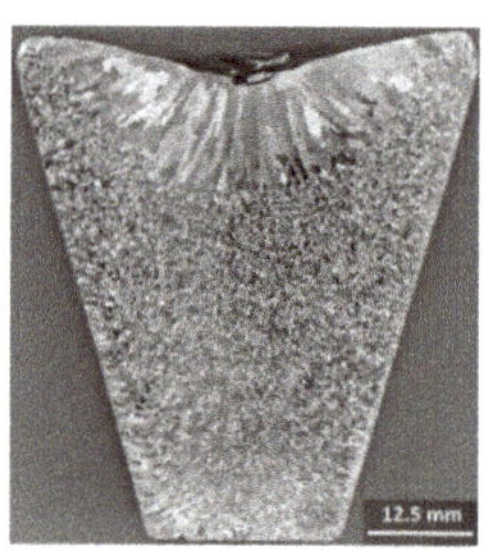

(a)

(b)

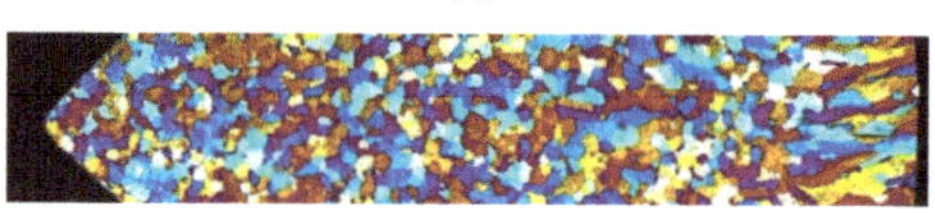

(c)

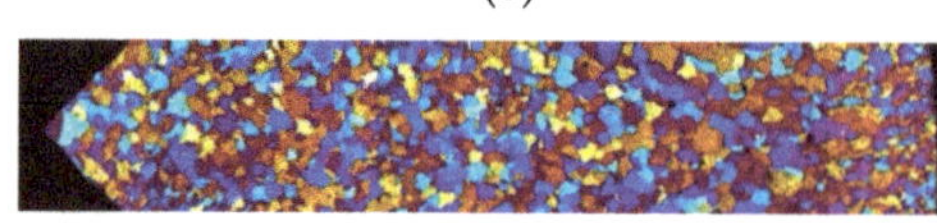

(d)

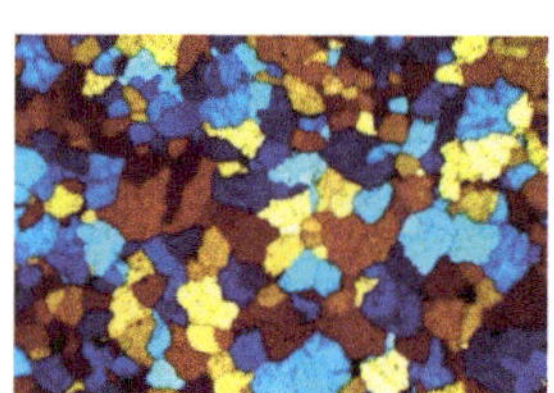

(e)

Figure 3. Effect of rod introduction into the melt in the temperature range 720 to 670 °C (**a**,**c**) and 710 to 670 °C (**b**–**e**) on the grain structure of the alloy cast in a TP1 mould.

Ultrasonic oscillations applied to the rod further refined the grain structure as illustrated in Figure 4. This might be due to the better distribution of crystal fragments in the melt as well as to the additional activation of insoluble inclusions by ultrasonic cavitation [13,17]. At the same time the grain size obtained upon rod introduction with ultrasonic oscillations in a wider temperature range was finer than in a narrower temperature range, i.e., 123 µm and 165 µm, respectively. As ultrasonic oscillations, in addition to accelerated rod dissolution, may facilitate additional mechanisms of grain refinement such as activation of inclusions and fragmentation of solid crystals with their effect being time-dependent, therefore, the longer the ultrasonic cavitation and vibration works, the better the result [13]. Hence, a slight increase in the rod introduction time provided by using a wider temperature range is beneficial for the grain refinement.

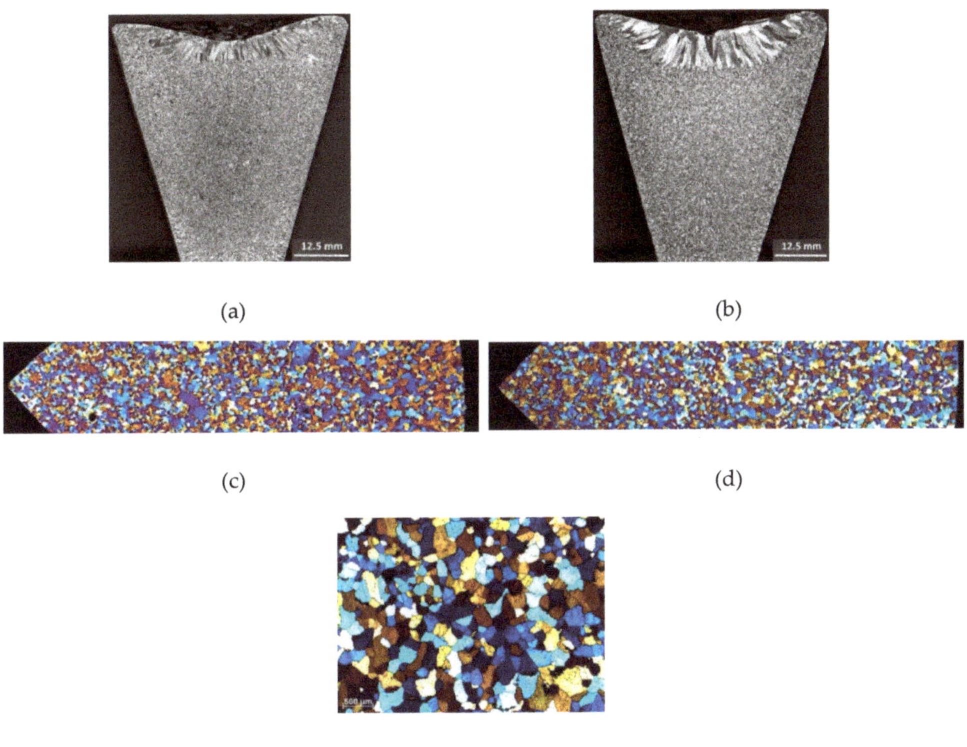

Figure 4. Effect of rod introduction with ultrasonic vibrations into the melt in the temperature range 720 to 670 °C (**a**,**c**) and 710 to 670 °C (**b**–**e**) on the grain structure of the alloy cast in a TP1 mould.

The standard procedure of grain refinement of aluminium alloys is the addition of an AlTiB master alloy. In order to compare the grain refining effects that we have achieved with the same-alloy rod introduction, we did additional experiments with the standard Al5Ti1B grain refining master alloy addition (0.2% of master alloy). Two experiments were performed, i.e., addition of the grain refiner to the base alloy and same-alloy rod introduction into the grain refined master alloy. The obtained grain sizes were rather small (112–123 µm), Figure 5, and comparable with those obtained upon same-alloy rod introduction with ultrasonic oscillations, Figure 4. Although the addition of a rod to the grain refined alloy gave a slight additional decrease in the grain size (112 µm vs 123 µm), it was not statistically significant. These results demonstrate that under given cooling conditions (TP1 mould) the addition of 0.2% grain refiner as well as the addition of the same-alloy rod gives the maximum number of solidification sites that can be realised and produce grains. Therefore the achieved grain size 112–123 μm is the smallest possible and does not depend on the way how the active solidification sites are introduced.

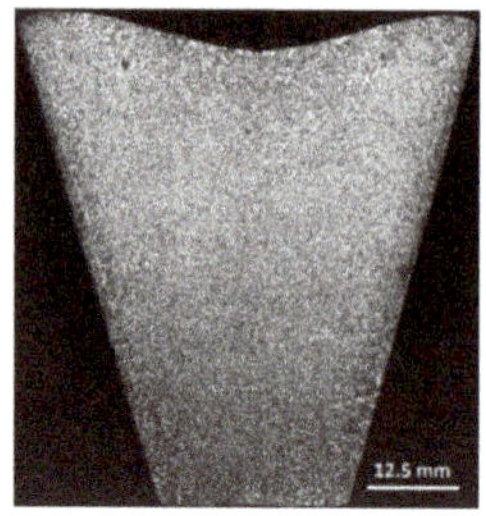

(a)

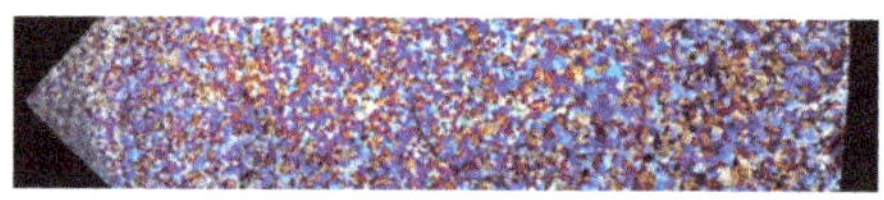

(b)

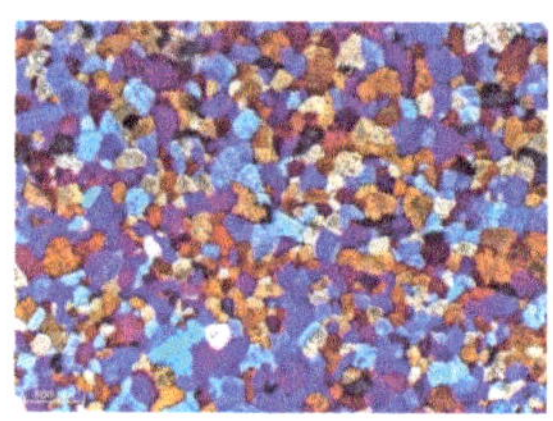

(c)

Figure 5. Effect of Al5Ti1B grain refiner addition and same-alloy rod introduction into the melt in the temperature range 710 to 670 °C on the grain structure of the alloy cast in a TP1 mould: (**a**), macrostructure; (**b**,**c**) microstructure.

Figure 6 summarises the results of grain size measurements for all studied cases.

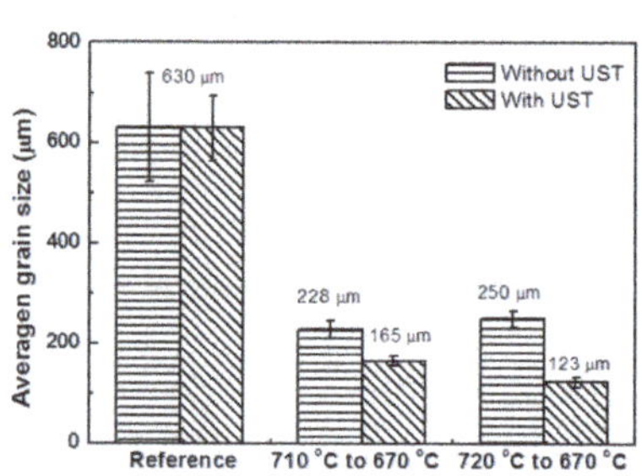

(a)

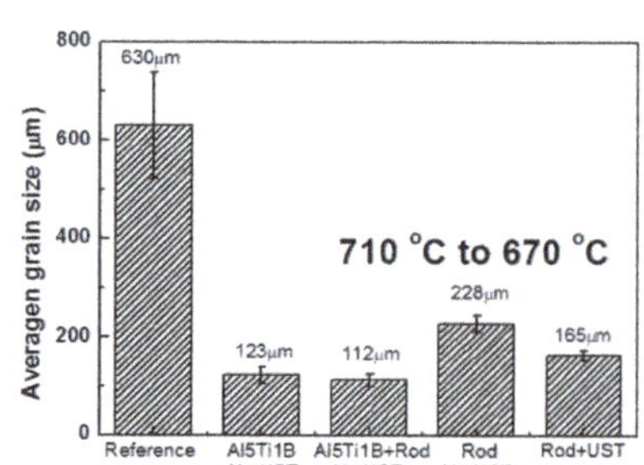

(b)

Figure 6. Grain sizes achieved in experiments: (**a**) the effects of the same-rod introduction, ultrasonic oscillations and the temperature range of introduction and (**b**) comparison of the same-alloy rod introduction with the standard grain refinement by master alloy addition.

This work has a limited scope of demonstrating the grain refining effect of the introduction of a same-alloy solid material into the melt facilitated by ultrasonic vibration of this rod. We based our discussion on the known mechanisms of suspension casting that include the introduction of active substrates and rapid undercooling of the melt [2–5] and known mechanisms of ultrasonic melt processing that include (in relation to the observed effects) accelerated dissolution and mixing [14] facilitated by fragmentation of solid suspended particles [13]. The combination of these effects produced a remarkable grain refinement effect, comparable with the standard procedure of grain refinement. The following research should focus on the further optimisation of the procedure with respect to the temperature range, processing time, amount of the introduced solid material and further cooling conditions during solidification. The effect of ultrasonic vibrations on the melt degassing on the one hand and on the introduction of potential oxides with the solid material on the other hand, should be further explored.

4. Conclusions

1. The grain refining effect upon introduction of a same-alloy rod into the melt prior to casting has been confirmed.
2. The temperature range of the same-alloy rod introduction should be close the casting temperature to assure that the generated solidification sites remain active.
3. The application of ultrasonic vibrations to the same-alloy rod gives additional grain refining effect that is most probably related to the known effects of cavitation on the accelerated dissolution and mixing as well as on the fragmentation of solidification sites.
4. The observed grain refining by the same-alloy rod introduction with the ultrasonic oscillations applied is similar to that achieved by the standard grain refining procedure.
5. The results show a promising technology of using the same-alloy rod instead of the grain refining rod in continuous casting processes.

Author Contributions: D.E. provided funding acquisition, supervision, conceptualization, methodology, participated in experiments, prepared the original draft and edited the final version. F.W. contributed to methodology, performed experiments and analysis and reviewed the draft paper.

Funding: The research was funded by the UK Engineering and Physical Science Research Council (EPSRC), Ultra-Cast project, grant number EP/L019884/1.

Acknowledgments: This paper is based on the presentation at the 16th International Conference on Aluminum Alloys (ICAA16, Montreal, Canada, 17–21 June, 2018. Technical assistance of S. Venditti and P. Lloyd in conducting experiments is highly appreciated.

Conflicts of Interest: The authors declare no conflict of interest.

References

1. Danilov, V.I.; Neimark, V.E. Effects of solidification conditions on the structure of an aluminium ingot. *Metallurg* **1938**, *10*, 34–43.
2. Scheil, E. Inoculation and inoculants in metal melts. *Giesserei* **1956**, *43*, 233–240.
3. Madyanov, A.M. *Suspension Casting*; Metallurgizdat: Moscow, Russia, 1969.
4. Zatulovsky, S.S. *Suspension Casting*; Naukova Dumka: Kiev, Russia, 1981.
5. Balandin, G.F. *Formation of Grain Structure in Castings*; Mashinostroenie: Moscow, Russia, 1973; pp. 225–227.
6. Schmidt, W. Method of Solidifying Molten Metal. U.S. Patent US3247557 (A), 26 April 1966.
7. Talbot, A.M.; Soller, H.W. Method of Internal Nucleation of a Casting. U.S. Patent GB1324526 (A), 25 July 1973.
8. Krupp GmbH. Foundry Mould Castings—With Solid Metal Particles Added to Molten Metal in Mould. U.S. Patent FR2201149 (A1), 26 April 1974.
9. Bondarev, B.I.; Napalkov, V.I.; Tararayshkin, V.I. *Modification of Aluminium Wrought Alloys*; Metallurgiya: Moscow, Russia, 1979; pp. 96–98.
10. Eskin, G.I. *Ultrasonic Treatment of Molten Aluminium*; Metallurgiya: Moscow, Russia, 1965.

11. Abramov, O.V. *Crystallization of Metals in Ultrasonic Field*; Metallurgiya: Moscow, Russia, 1972.
12. Abramov, O. *High Intensity Ultrasonics. Theory and Industrial Application*; Gordon and Breach: Amsterdam, The Netherlands, 1998.
13. Eskin, G.I.; Eskin, D.G. *Ultrasonic Treatment of Light Alloy Melts*, 2nd ed.; CRC Press: Boca Raton, FL, USA, 2015.
14. Brodova, I.G.; Popel, P.S.; Eskin, G.I. *Liquid Metal Processing. Application to Aluminum Alloy Production*; Taylor and Francis: London, UK, 2002.
15. Komarov, S.; Oda, K.; Ishiwata, Y.; Dezhkunov, N. Characterization of acoustic cavitation in water and molten aluminum alloy. *Ultrason. Sonochem.* **2012**, *20*, 754–761. [CrossRef] [PubMed]
16. Aluminum Association. *Standard Test Procedure for Aluminum Grain Refiners, TP-1*; The Aluminum Association: Washington, DC, USA, 1990.
17. Wang, F.; Eskin, D.; Connolley, T.; Wang, C.; Koe, B.; King, A.; Reinhard, C.; Mi, J. In-situ synchrotron X-ray radiography observation of primary Al_2Cu intermetallic growth on fragments of aluminium oxide film. *Mater. Lett.* **2018**, *213*, 303–305. [CrossRef]

metals

MDPI

Article

Effect of a Traveling Magnetic Field on Micropore Formation in Al-Cu Alloys

Yanjin Xu [1,*], Lijun Wei [2], Baoshuai Han [1], Enyu Guo [3,*], Mingyue Wang [2] and Yanqing Su [4]

1 AVIC Manufacturing Technology Institute, Beijing 100024, China; hbshit@126.com
2 School of Material Science and Engineering, Beihang University, Beijing 100191, China; bhwlj@buaa.edu.cn (L.W.); bhwangmy@163.com (M.W.)
3 Key Laboratory of Solidification Control and Digital Preparation Technology (Liaoning Province), School of Materials Science and Engineering, Dalian University of Technology, Dalian 116024, China
4 School of Materials Science and Engineering, Harbin Institute of Technology, Harbin 150001, China; suyq@hit.edu.cn
* Correspondence: xuyj_avic@aliyun.com (Y.X.); eyguo@dlut.edu.cn (E.G.); Tel.: +86-10-8570-1653 (Y.X.); +86-411-8470-9500 (E.G.)

Received: 14 May 2018; Accepted: 5 June 2018; Published: 12 June 2018

Abstract: The effect of traveling magnetic fields (TMFs) on the grain and micro-pore formation in an Al alloy was studied by scanning electron microscope and X-ray microtomography in this work. The results show that with the increasing magnetic flux density, the three-dimensional morphology of the micro-pores transformed from dendrite to a relatively equiaxed structure. Quantified results show that both the micro-pore volume fraction and the average grain size of the primary phase decreased as the TMF density increased. The analyses show that the forced convection induced by TMF can break the dendrites, refine the grain size, and promote the liquid feeding, leading to the decrease in the volume fraction of the porosity and improved mechanical property. The TMF performed at different stages during solidification reveal that the maximum effect of TMF on reducing the micro-pore formation was found when TMF was applied in the stage of nucleation and the early stage of grain growth during solidification.

Keywords: aluminum alloy; solidification; X-ray tomography; porosity defects; magnetic field treatment

1. Introduction

The shape casting process offers an effective way to produce complex components in one single production step [1]. However, the use of this technique is also limited by the formation of defects such as micro-pore, hot tearing [2,3], and so forth. Porosity/micro-pore is one of the major defects in castings, which is usually induced by gas segregation and solidification shrinkage in the mushy zone. According to the research of Campbell [4], the presence of micro-pore reduces the mechanical properties of the cast component, including fatigue life, tensile strength, ductility, and surface quality.

No porosity could be found in the castings if the gas is absent and the feeding is adequate. However, many regions of the castings are not fed, and then the micro-pore may form in a number of ways [4]. The problem of micro-pore formation in casting alloys continues to be of interest despite the many computational models that have been proposed. The problem of the micro-pore formation is very complex, as it involves many materials (that is, initial gas content, melt purity) and processing (that is, temperature, temperature gradient, cooling rate, applied pressure) parameter interactions in complex physics [5]. It has been shown that the nucleation of the microporosity is influenced by foreign impurities [6]. According to the previous work performed by Felberbaum [7], we know that the porosity which is constrained to grow between narrow inter-dendritic liquid channels has a higher curvature, and thus a higher internal pressure than that of a free-growth spherical one, and the fraction of the porosity, hence, decreases with an increasing curvature. Therefore, increasing the curvature of the porosity and decreasing the grain size are effective methods to decrease the fraction of the porosity.

The introduction of the electromagnetic technique as a new method for tailoring the microstructure and micro defection of alloys has attracted increasing attention. The application of the high magnetic field suggested a possibility of adjusting the morphology of solid-liquid phase during the solidification of Al-Cu alloys [8], which shows a potential way to control the distribution and amount of porosities. Previous research [1,2] found that the application of the magnetic field can cause grain refinement, in which some of the resulting microstructures are much better than those used by other solidification technology, for example, supergravity solidification [9]. Zuo et al. [10] pointed out that the increasing external magnetic field can tilt the growth direction of the lamellar eutectic and decrease the coarse eutectic lamellar spacing during solidification, leading to a higher strength. The work by Erb et al. [11] indicated that even a low magnetic field (1 to 10 millitesl as) has a significant effect on the orientation and distribution of the reinforced particles in artificial composites. Therefore, the application of a magnetic field during the solidification process is an effective method to reduce defects and optimize microstructure and properties. However, despite the successful applications of the magnetic field in the material fabrication, the mechanisms of the effect of the electromagnetism on the melt are not yet well understood. Therefore, more work is needed to understand the role of the magnetic field in each stage of the solidification progress.

The recent development of high-resolution X-ray tomography imaging techniques proposed a useful method to explore the structure evolution in the solidification process. Holm et al. [12] have enabled three-dimensional (3D) observations of the microporosity morphology. Lee and Hunt [13] first applied this technique to visualize the formation of the porosity in Al-Cu alloys with a micro-focus X-ray source. However, the resolution was about 25 μm and the intensity of the beam was also a limiting factor. The synchrotron X-ray tomography provides higher resolution and higher flux capabilities at beamlines, and, thus, better characterization can be performed, which results in an improved understanding of the mechanisms involved in the material processing [14].

In this work, the effect of TMF on the micro-pore formation and on the morphology of micro-pores and grain size are compared in the solidification process of Al-Cu alloys under a traveling magnetic field (TMF) with various magnetic flux densities. X-ray tomography was performed to characterize the 3D morphology of the micro-pores in Al-Cu alloys after the alloy was solidified. Meanwhile, different strategies of magnetic treatments were performed in different stages of solidification to evaluate the effect of this treatment, and its mechanism was discussed and revealed.

2. Materials and Methods

The schematic illustration of the experimental apparatus is shown in Figure 1. The traveling magnetic field was created by means of applying out-of-phase alternating current (AC) currents (50 Hz) to six coils with a star arrangement connected vertically one upon the other. As a result, a downward meridional traveling Lorentz magnetic field was induced in the sample melt. This technique has already been detailed in previous work [15]. A nonmagnetic stainless-steel frame was designed and assembled to hold the sample firmly. The sample had a diameter of 3.5 mm and a length of 100 mm and was placed in a quartz tube fixed in the center of the cavity. The nonmagnetic stainless-steel frame, which allowed water to stream in and out, was installed and thus enabled the equipment to control the cooling rate by adjusting the flow rate of the streaming water. An electric heating furnace was used to heat and melt the sample. The furnace was removed through the opening of the upper cylinder when the TMF generator was applied.

The Al alloy studied in this work had a composition of 3.8Cu-0.4Mn-0.3Ti-0.1Cd-0.25V-Al (bal. in wt. %). The sample was placed in the center of the TMF generator and the heating furnace was set around the sample and turned on to heat and melt the sample up to 1023 K. The molten sample was kept at this temperature for 10 min before the furnace was removed. Then the TMF generator was applied at different stages of the cooling process to the ambient temperature. The magnetic flux densities of the TMF were set as 24 mT, 32 mT, and 48 mT, respectively. For comparison, a sample was also prepared without TMF application.

The microstructures of the as-cast samples were examined by scanning electron microscopy (SEM) (Quanta 200FEG, FEI, Hillsborough, OR, USA) in the cross-section of the sample and 3D X-ray microtomography. The X-ray microtomography was carried out at the BL13W1 beamline in the Shanghai Synchrotron Radiation Facility (SSRF). An X-ray energy of 30 keV was used to penetrate a cylindrical sample of 3.5 mm in diameter and 2–5 mm in height with a voxel size of 3.7 μm. A 2048 × 2048 pixel CoolSnap K4 CCD camera (Coolsnap, Photometrics, Tucson, AZ, USA), coupled with a $CdWO_4$ scintillator screen, was used to record the projections. Each 3D data set contains1440 projections that were collected with an exposure time of 1 s. The projections were reconstructed using a filtered-back-projection algorithm. The raw images were preprocessed in Image J (NIH, Bethesda, MD, USA) and then further analyzed in Avizo (VSG Group, Bordeaus, France). A median filter was applied in Avizo to reduce the noise of the image. The tensile tests were conducted on an electronic universal mechanical testing machine (Instron5569, Instron Co., Canton, OH, USA) at an initial strain rate of 5×10^{-2}. The length of the sample gauge was 20 mm.

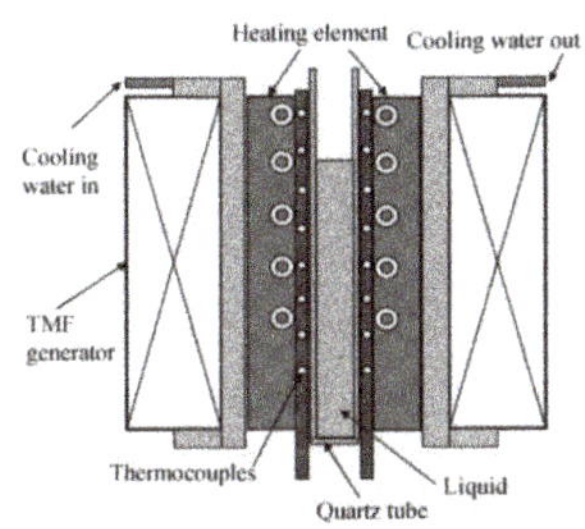

Figure 1. The schematic illustration of the experimental apparatus.

3. Results and Discussion

In order to determine the effect of magnetic flux density on the volume fraction and morphology of the micro-pores, Al-alloy samples were solidified in the TMF with various magnetic flux densities. Figure 2 shows the typical microstructures formed without TMF treatment (Figure 2a) and with 48 mT TMF treatment (Figure 2b). The microstructure was composed of primary dendrites and white Al-Al_2Cu eutectic phase. It is evident that micro-pores (dark regions) were present in the interstices between the primary dendrites in the sample without TMF treatment, and shrinkage porosity appears as isolated holes, both of which indicated that feeding without TMF treatment in the densely packed dendrite arms was very difficult. For comparison, the micro-pores were inhibited significantly after TMF treatment, as shown in Figure 2b, suggesting that the TMF has a significant effect on reducing the porosity defects.

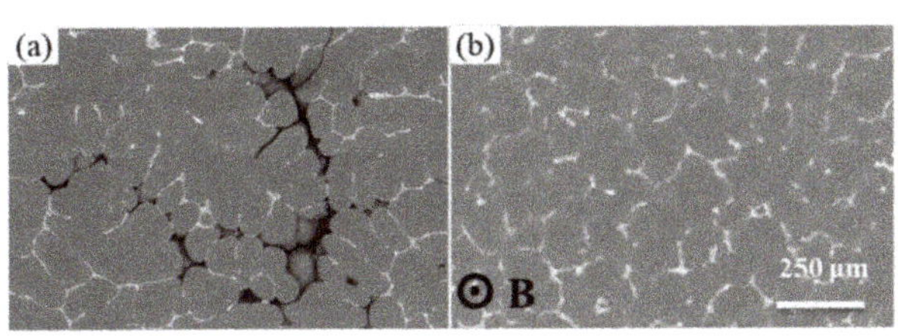

Figure 2. The SEM images showing microstructures in the as-cast Al alloys: (**a**) without and (**b**) with 48 mT TMF treatment. Shrinkage porosities (dark regions) are observed in the Al-alloy without TMF treatment. The direction of the TMF is indicated in (**b**) by point B, represents from inward to outside. Scales bar is the same for (**a**,**b**).

The 3D shapes of the typical micro-pores formed in Al alloys treated under various magnetic flux densities are shown in Figure 3. Without TMF treatment, shown in Figure 3a, the micro-pores have an elongated and highly-connected shape along the thermal gradient direction (horizontal). However, with the application of TMF, the morphologies of the micro-pores are more likely to be equiaxed, and the distribution of the micro-pores has been changed significantly. Meanwhile, the size of the micro-pores tends to decrease gradually with the increase of the magnetic flux density.

A schematic showing the effect of TMF on the distribution and morphology of the micro-pores is presented in Figure 3e. Due to the low copper content, the volume fraction of the primary phase was very high (~90%) in the solidification microstructure. When the volume fraction of the primary phase is low in the early solidification progress, a significant amount of fluid can flow through the dendrite network and feed the solidification shrinkage. As the solidification progress continues, the dendrite would grow and interconnect to form a complex extended network in the three-dimensional space, as can be seen in Figure 3. In this situation, the micro-pores have to nucleate in the mushy region after the formation of the primary aluminum dendrites, and then grew by replacing and pushing away the surrounding liquid. As the pushing force induced by the pore was not high enough to break the primary phase, the solid phase grows along or even into the micro-pores. As a result, the micro-pores formed in this way had a very complex morphology. In the case of the alloy in this work, the solidification temperature range is ~120 °C, and the fraction of the eutectic microstructure is relatively small. This makes the feeding very difficult for the rest of the liquid to solidify, which usually results in high levels of micro-pores. Since the micro-pores have formed in the mushy zone after the solidification of the primary phase "skeleton", the shape of the micro-pores is determined by the morphology of the primary phase accordingly. It was observed that micro-pores forming near the dendritic phase were also dendritic in structure.

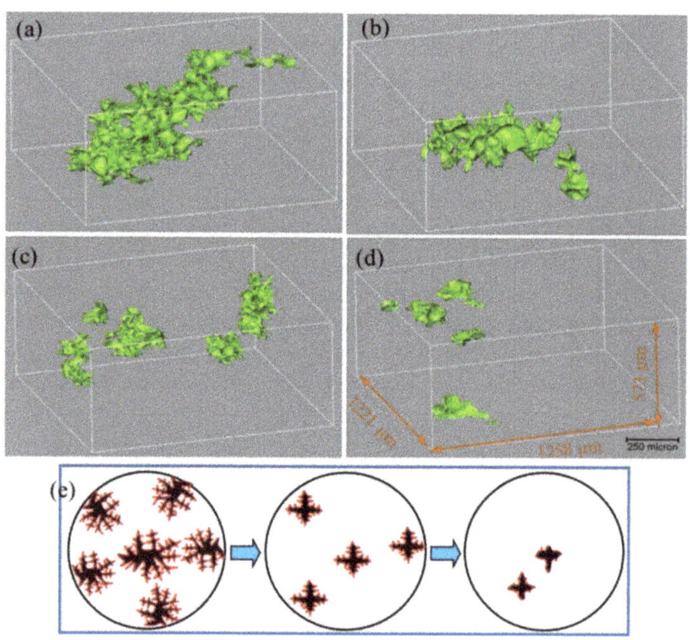

Figure 3. The 3D morphology of the micro-pores in the Al-Cu alloys under TMF with a magnetic flux density of (**a**) 0 mT (**b**) 24 mT (**c**) 32 mT (**d**) 48 mT. (**e**) is the schematic showing the effect of TMF on the distribution and morphology of the micro-pores.

In order to better understand and evaluate the contribution of TMF, a special study was designed to characterize and establish the relationship between the quantified micro-pore fraction and grain size under various experimental conditions, and the result is shown in Figure 4. It is found that the volume fraction of micro-pore decreases gradually with the increasing TMF intensity and decreasing grain size. In the case of our study, TMF plays important roles in two ways: one is that it refines the grains and the other one is that it enhances the feeding pressure [16,17]. In theory, grain refinement can promote mass feeding [3]. Practically, when the solid fraction exceeds a threshold value, the dendrites start to become a coherent network and the free grains are so large that only smaller ones can pass through the network, thus, leaving the large grains pinned in the solid skeleton and the feeding path blocked prematurely. When TMF is introduced, the solid skeleton is likely to be broken up by the imposed Lorentz force during solidification, and the grains are refined [18]. Moreover, due to the stirring effects of the TMF, the distribution of solutes becomes more uniform, which are helpful to precipitate equiaxed grains instead of the solid skeleton [19]. Therefore, the slurry of the liquid metal can move to the front of the solidification interface sufficiently, increasing feeding ability and decreasing the volume fraction. Thus, it is assumed that the suppression of the shrinkage porosity under TMF was mainly caused by different aspects including the grain refinement, the enhanced feeding pressure induced by TMF and a combined effect of them. With the increase of the TMF intensity, the grains were refined gradually and became more equiaxed, making the micro-pores transform from dendrite to equiaxed structure accordingly, as shown in Figure 3b–d. These results indicate that the TMF can change the micro-pore morphology, and this change is mainly caused by the shape of the primary solid skeleton. Although many research reports [14,18,20] have studied the grain refining under the magnetic field, the relationship between the grain refining and porosity has not been revealed. In this work, we found that refining the grain size can improve the feeding ability in the solidification progress and decrease the volume fraction of the porosity under TMF, suggesting a promising process to improve the quality of the casting Al-alloy.

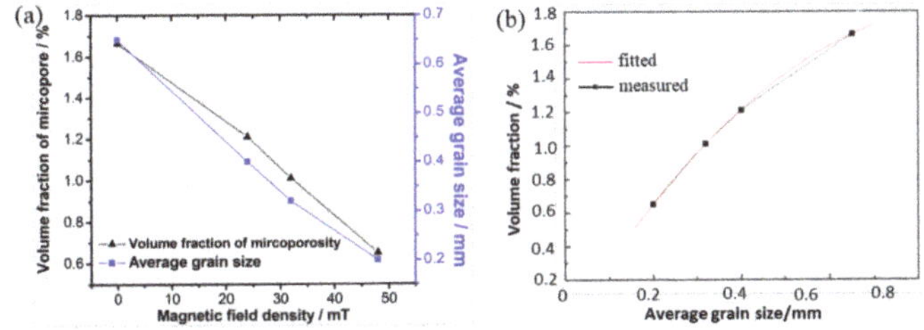

Figure 4. (**a**) The effect of TMF on the micro-pore fraction and grain size in Al-alloy solidified under different magnetic flux conditions; (**b**) result showing the relationship between the volume fraction of the micro-pore and the average primary grain size.

The cooling curves were recorded during all the experiments and the influence of TMF on the cooling curve was found to be insignificant. Thus, the TMF feeding mechanism is discussed based on its effect on different stages. In order to elucidate the feeding mechanism of the shrinkage porosity formation of this material under TMF, nine treatment strategies under the TMF intensity of 48 mT were carried out according to the cooling curve of the specimen solidified without the TMF, as shown in Figure 5a. The relationship between the volume fraction of micro-pores and the grain size with a given TMF treatment at different solidification stages is presented in Figure 5b, which indicates the specific stage where the TMF played the main role. The TMF treatment at Stage 1 and 2, where the alloy was in a complete liquid status, shows no effect on either grain refinement or micro-pore fraction. This indicates that the TMF has no inoculation effect on the liquid Al alloy. When the TMF was applied in the nucleation Stage 3 (before crystal growth had started), both the grain size and the volume fraction of the micro-pores was reduced, as compared with those in Stage 1 and 2, suggesting that the

TMF can refine grains and suppress the shrinkage porosity by imposing effects on the nucleation stage. Our results are in agreement with the report by Liao et al. [21]. The results also show that the TMF in Stage 3 only affected the grain nucleation partly, while the TMF applied in the whole nucleation stage and the early stage of the grain growth, that is, Stage 4 and 5 respectively, can further play an important role in reducing the grain size. Both the strategies led the similar results, yielding refined grains and suppressed shrinkage porosity. Moreover, both the volume fraction of the micro-pores and the average grain size are minimal with the TMF treated at these two stages. At Stage 6, TMF was introduced within the middle crystal growth stage, wherein micro-pores were suppressed slightly due to grain refinement. It can also be seen from Figure 5b that the application of the TMF at the last stage of crystal growth almost has no effect on either the grain size or the micro-pore volume fraction (Stages 7 and 8). Interestingly, when applying the TMF for the entire solidification process (Stage 9), the grains were refined greatly and the volume fraction of the micro-pores was also significantly reduced, similar to that observed in Stage 4 and 5. Therefore, it is concluded that the defect of the pore formation and the mechanism of grain refinement are closely related to the nucleation process and that, in order to maximize the grain refinement effect, TMF should be applied in Stage 4 and 5 or the whole process stage. Consider the resources, that is, the time, efforts, and the instrument time for the TMF generation, applying TMF at Stages 4 and 5 should be enough instead of applying it for all the stages.

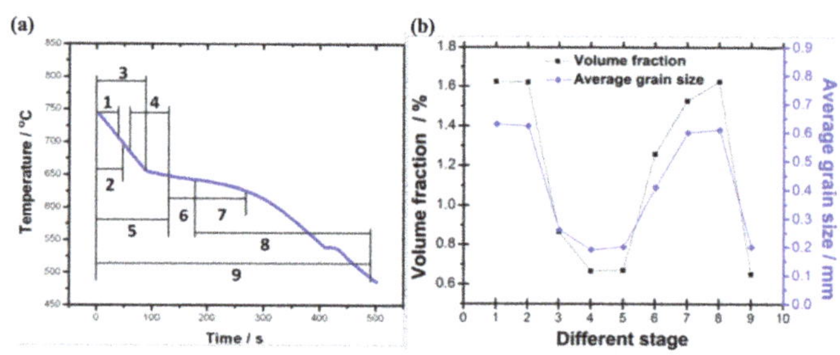

Figure 5. The effect of TMF on the micro-pore fraction and grain size in Al-alloy solidified under different conditions: (**a**) schematic sketch of solidification stages treated with TMF; (**b**) evolution of volume fraction and average grain size in different solidification stages with TMF treatment.

The aforementioned experimental results clearly show that grain refinement by TMF is primarily achieved during the nucleation stage and the early stage of the crystal growth other than in the stages when only the liquid phase exists or in the last stage of crystal growth. Shrinkage porosity suppression, accompanied by grain refinement, is achieved at the same stage under TMF application. Compared with the two curves in Figure 5b, it is obvious that the change of the volume fraction of the micro-pores exhibits the same trend as the grain size, which supports our hypothesis that grain refinement is helpful to suppress the porosity. According to the research of Fleming [22], we know that the shear strength of the solid phase can be neglected when the volume fraction of the solid phase is below 0.2 and that this shear strength increases with the increase of the solid phase volume fraction when the fraction is larger than 0.2. Therefore, the forced convection induced by TMF can break the small dendrites easily in the nucleation progress, that is, Stage 3, 4, and 5, which leads to a great refining. However, the shear strength of the solid phase increased dramatically after the fraction of the solid phase reached 0.4. In our study, the electromagnetic force induced by TMF is relatively low. It is believed that the forced convection induced by TMF cannot break the dendrites in this situation. This might account for the observation that TMF cannot refine the grains in the last stage of crystal growth. In the last crystal growth stage, although some residual liquid still existed in the interdendritic zone, the micro-pore formation cannot be suppressed even with TMF. This suggests that the enhanced electromagnetic pressure has no effect on the shrinkage porosities. As a result, the most appropriate stage to apply the TMF is in the nucleation stage and the early stage of the grain growth when the solid phase is below 0.2, in which the TMF can refine the grain to the greatest extent in this process and suppress the porosity in the solidification.

Due to the decreased micro-pores, the mechanical properties of the alloys under TMF were enhanced and the tensile results are shown in Figure 6. For example, the tensile strength and elongation of the alloy increased with the increasing TMF intensity. The optimal mechanical properties (strength of 214 MPa, elongation of 11.3%, increased by 8.0% and 29.9%, respectively, as compared with the samples without TMF treatment) were obtained in the Al alloy treated with 48 mT TMF. Under a tensile loading condition, cracks would be initiated from the weak position, that is, micro-pores, resulting in an early failure of the sample when a number of pores are present in the material. The small grains present in the materials also leads to a greater strength and better elongation. This also suggests that the effect of TMF has a positive effect on improving the mechanical properties. Therefore, TMF treating is a promising method to improve the microstructure and mechanical properties of the casting Al alloys.

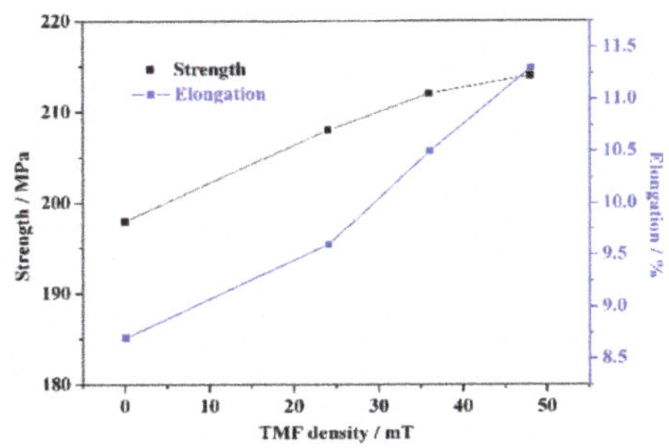

Figure 6. The tensile strength and elongation of Al-alloys solidified under TMF.

4. Conclusions

In summary, we studied the effect of a magnetic field on the micro-pore formation in an Al alloy. It is found that TMF treatment can suppress the shrinkage porosity effectively, resulting in a reduction in both volume fraction of the micro-pore and grain size after TMF treatment. The analyses show that TMF can enhance the feeding ability in the solidification progress, resulting in decreased volume fraction of the porosity in the Al alloy. The grain refinement can also be introduced by the TMF. Detailed mechanism investigations revealed that micro-pores can only be suppressed by the TMF when it is applied during nucleation and the early stage of the grain growth. Therefore, it is suggested that the TMF be applied in the nucleation stage instead of the whole solidification progress. The refined grains and reduced micro-pores caused by the TMF in the materials lead to improved mechanical properties.

Author Contributions: Y.X. guided the investigation and experiments, analyzed the experiment data and mechanism, wrote the paper, and revised the paper; L.W. optimized the research plan and revised the paper; B.H. made the diagram and graph, and revised the paper; E.G. made the diagram and graph, and revised the paper; M.W. collected the data, and revised the paper; Y.S. searched literature, and optimized the diagram and charts.

Funding: This research was funded by the Defense Industrial Technology Development Program (Grant No. JCKY2017205B032).

Acknowledgments: The use of X-ray synchrotron beam line BL13W1 at the Shanghai Synchrotron Radiation Facility (SSRF) is acknowledged. The authors wish to express thanks to them. Helpful discussion with James Mertens in the Arizona State University is also greatly appreciated.

Conflicts of Interest: The authors declare no conflict of interest.

References

1. Meidani, H.; Desbiolles, J.L.; Jacot, A.; Rappaz, M. Three-dimensional phase-field simulation of micropore formation during solidification: Morphological analysis and pinching effect. *Acta Mater.* **2012**, *60*, 2518–2527. [CrossRef]
2. Dantzig, J.A.; Rappaz, M. *Solidification*; EPFL: Lausanne, Switzerland, 2009; pp. 479–500.
3. Liotti, E.; Lui, A.; Vincent, R.; Kumar, S.; Guo, Z.; Connolley, T.; Dolbnya, I.P.; Hart, M.; Arnberg, L.; Mathiesen, R.H. A synchrotron X-ray radiography study of dendrite fragmentation induced by a pulsed electromagnetic field in an Al-15Cu alloy. *Acta Mater.* **2014**, *70*, 228–239. [CrossRef]
4. Campbell, J. *Castings*; Butterworth-Heinemann: London, UK, 2003; pp. 280–396.
5. Stefanescu, D.M.; Catalina, A.V. Physics of microporosity formation in casting alloys—Sensitivity analysis for Al-Si alloys. *Int. J. Cast Met. Res.* **2011**, *24*, 144–150. [CrossRef]
6. Roy, N.; Samuel, A.M.; Samuel, F.H. Porosity formation in Al-9 Wt pct Si-3 Wt pct Cu alloy systems: Metallographic observations. *Metall. Mater. Trans. A* **1996**, *27*, 415–429. [CrossRef]
7. Felberbaum, M.; Rappaz, M. Curvature of micropores in Al-Cu alloys: An X-ray tomography study. *Acta Mater.* **2011**, *59*, 6849–6860. [CrossRef]
8. Li, X.; Fautrelle, Y.; Ren, Z. Influence of an axial high magnetic field on the liquid-solid transformation in Al-Cu hypoeutectic alloys and on the microstructure of the solid. *Acta Mater.* **2007**, *55*, 1377–1386. [CrossRef]
9. Yang, Y.; Song, B.; Yang, Z.; Song, G.; Cai, Z.; Guo, Z. The refining mechanism of super gravity on the solidification structure of al-cu alloys. *Materials* **2016**, *9*, 1001. [CrossRef] [PubMed]
10. Zuo, X.; Zhao, C.; Zhang, L.; Wang, E. Influence of growth rate and magnetic field on microstructure and properties of directionally solidified Ag-Cu eutectic alloy. *Materials* **2016**, *9*, 569. [CrossRef] [PubMed]
11. Erb, R.M.; Libanori, R.; Rothfuchs, N.; Studart, A.R. Composites reinforced in three dimensions by using low magnetic fields. *Science* **2012**, *335*, 199–204. [CrossRef] [PubMed]
12. Holm, E.A.; Duxbury, P.M. Three-dimensional materials science. *Scr. Mater.* **2006**, *54*, 1035–1040. [CrossRef]
13. Lee, P.D.; Hunt, J.D. Hydrogen porosity in directional solidified aluminium-copper alloys: In situ observation. *Acta Mater.* **1997**, *45*, 4155–4169. [CrossRef]
14. Zhang, Y.; Song, C.; Zhu, L.; Zheng, H.; Zhong, H.; Han, Q.; Zhai, Q. Influence of electric-current pulse treatment on the formation of regular eutectic morphology in an Al-Si eutectic alloy. *Metall. Mater. Trans. B* **2011**, *42*, 604–611. [CrossRef]
15. Su, Y.; Xu, Y.; Zhao, L.; Guo, J.; Fu, H. Effect of electromagnetic force on melt induced by traveling magnetic field. *Trans. Nonferrous Met. Soc. China* **2010**, *20*, 662–667. [CrossRef]
16. Xu, Y.; Su, Y.; Luo, L.; Liu, J.; Guo, J.; Fu, H. Study on in situ Al-Si functionally graded materials produced by traveling magnetic field. *Sci. Eng. Compos. Mater.* **2012**, *19*, 209–214. [CrossRef]
17. Su, Y.; Xu, Y.; Luo, L.; Zhang, T.; Wu, S.; Guo, J. Study on traveling magnetic field casting of sheet components. *China Foundry* **2010**, *7*, 259–264.
18. Zhang, Y.; Cheng, X.; Zhong, H.; Xu, Z.; Li, L.; Gong, Y.; Miao, X.; Song, C.; Zhai, Q. Comparative study on the grain refinement of Al-Si alloy solidified under the impact of pulsed electric current and travelling magnetic field. *Metals* **2016**, *6*, 170. [CrossRef]
19. Zhang, C.; Zhong, H.G.; Wu, C.S.; Ni, J.; Zhai, Q.J. Effect of cooling rate and forced convection on as-cast structure of 2205 duplex stainless steel. *China Foundry* **2015**, *12*, 32–38.
20. Metan, V.; Eigenfeld, K.; Räbiger, D.; Leonhardt, M.; Eckert, S. Grain size control in Al-Si alloys by grain refinement and electromagnetic stirring. *J. Alloy Compd.* **2009**, *487*, 163–172. [CrossRef]
21. Liao, X.; Zhai, Q.; Luo, J.; Chen, W.; Gong, Y. Refining mechanism of the electric current pulse on the solidification structure of pure aluminum. *Acta Mater.* **2007**, *55*, 3103–3109. [CrossRef]
22. Flemings, M.C. Behavior of metal alloys in the semi-solid. *Mater. Trans. A* **1991**, *22*, 957–981. [CrossRef]

metals

MDPI

Communication

Role of Roll Separating Force in High-Speed Twin-Roll Casting of Aluminum Alloys

Min-Seok Kim [1,*], Hyoung-Wook Kim [1], Su-Hyeon Kim [1] and Shinji Kumai [2]

1 Aluminum Department, Korea Institute of Materials Science, 797 Changwondae-ro, Seongsan-gu, Changwon, Gyeongnam 51508, Korea; hwkim@kims.re.kr (H.-W.K); shawnkim@kims.re.kr (S.-H.K.)
2 Department of Materials Science and Engineering, Tokyo Institute of Technology, 2-12-1 O-okayama, Meguro-ku, Tokyo 152-8552, Japan; kumai.s.aa@m.titech.ac.jp
* Correspondence: mskim@kims.re.kr; Tel.: +82-55-280-3397; Fax: +82-55-280-3599

Received: 3 May 2019; Accepted: 1 June 2019; Published: 3 June 2019

Abstract: The role of the roll separating force in the high-speed twin-roll casting of aluminum alloys was examined. In horizontal-type twin-roll casting, as the casting speed increased upon decreasing the roll separating force, the strip texture changed from a shear and rolling texture to a random texture. Direct temperature measurements during high-speed twin-roll casting showed that the roll separating force played a significant role in maintaining a good contact between the strip and the roll surface. This resulted in a high cooling rate around the roll nip and enabled the fabrication of a sound strip with a fine microstructure. Moreover, the high casting speed and lowered roll separating force gave a band structure consisting of fine globular grains in the mid-thickness region of the strip, which could be considered beneficial in the formation of a well-dispersed center segregation.

Keywords: high-speed twin-roll casting; roll separating force; Al-Si alloy; Al-Mg alloy

1. Introduction

In aluminum sheet production, the twin-roll casting (TRC) process has attracted increasing attention due to its economic advantages when compared with conventional direct-chill slab casting and hot rolling processes. Due to the rapid cooling rate of >100 K/s that is employed in the TRC process, the production of 3–10 mm thick strips can be achieved directly from the molten metal. The traditional TRC process was designed to combine the metal casting and hot rolling processes into a single operation, and in general mass production, a horizontal-type (or slightly tilted) caster is typically employed, where the casting speed is optimized to give a range of <2 m/min. However, high-speed twin-roll casting (HSTRC) has recently been developed using a vertical-type caster [1], with the primary aim being to improve the casting speed by rapid cooling rather than through applying the hot rolling effect. Therefore, copper rolls with relatively higher thermal conductivities compared to their steel roll counterparts were employed for this purpose, and the roll separating force (RSF) was reduced to one tenth of the level of the conventional horizontal-type TRC. These conditions allowed the casting speed to be increased to 180 m/min [2].

In the solidification stage of TRC, solid shells begin to grow on the roll surfaces after the molten metal is supplied between two rotating rolls. At a certain point, typically referred to as the "kiss point," two solidifying shells encounter one another, and joint shells are introduced to the roll gap to produce a single strip. In this process, the RSF is applied to the hot strip when the solidifying shells widen the roll gap. Since the strip is subjected to external stress during solidification, the RSF plays an important role in determining the cast strip microstructure [3].

Thus, the influence of the RSF on the cooling behavior and on the solidification structure of such strips is investigated in this study, and in particular in the high casting speed range. Through microstructural and textural observations, the role of the RSF is clarified, and the mechanism of strip formation is discussed.

2. Materials and Methods

Al-2Si, Al-7Si-0.3Mg, and Al-5Mg (wt%) alloys were melted in an electric furnace, and degassing was conducted using Ar gas for 10 min prior to casting. Lab-scale horizontal- and vertical-type twin-roll casters were used for strip fabrication. The horizontal-type caster consists of two Cu-Cr rolls, which are fixed firmly. Using this caster with minimized lubrication on the roll surface, the casting speed could be increased to 5 m/min. To compare the strip microstructure and texture, the steel roll caster in the mass-production line was also used to fabricate 1250 mm wide strips, which exhibited the typical microstructure of traditional TRC strips. For the vertical-type caster, pure Cu rolls with no lubrication on the roll surface were used to maximize the cooling rate, and one of the rolls was loosely supported by a series of springs. This condition makes it possible to maintain a low RSF at a high-speed casting range. The roll size of both lab-scale casters was 300 mm in diameter, and the rolls were internally cooled by running water during the strip casting process. The strip width fabricated by both casters was 100 mm. A direct temperature measurement technique [4] was adopted to investigate the cooling behavior of the strip during vertical-type HSTRC. A K-type thermocouple (ANBE SMT Co, Yokohama, Japan) was flown directly into the molten pool to measure the temperature from the center line of the upper side of the melt pool to the strip center. The casting speed was 60 m/min, and the RSF was varied from 3 to 60 kN. The temperature change was recorded every 0.5 ms using a data logger (NR600, KEYENCE, Osaka, Japan). For the cast strip, longitudinal cross-sectioned samples were mounted in epoxy resin prior to mechanical polishing. The polished samples were then etched using a 2% solution of Hydrogen fluoride (HF) in distilled water for microstructural observations using an optical microscope. The samples were also anodized at 40 V in a 3.3% solution of HBF_4 in distilled water to reveal their grain structures. The strip textures were observed by electron backscattering diffraction (EBSD) using a TESCAN MIRAII FE-SEM (Brno, Czech Republic) equipped with a Hikari EBSD detector at an accelerating voltage of 20 kV. The EBSD results were analyzed using OIM analysis 7 software provided by TSL, Co., Ltd. An automatic serial sectioning machine (UES Inc., Robo-Met. 3D, Dayton, OH, USA) and 3-D analysis software (FEI, Avizo Fire 7, Hillsboro, OR, USA) were used for three-dimensional characterization of the center segregation [3]. For the as-cast Al-5Mg strips, tensile tests along both the casting direction (CD) and the transverse direction (TD) were conducted using flat tensile specimens with a 25 mm gauge length and 6 mm gauge width. The fracture surfaces were observed using scanning electron microscopy (SEM; JSM-6610LV, JEOL, Tokyo, Japan).

3. Results and Discussion

3.1. Increasing the Casting Speed in the Horizontal-Type TRC Process

To maximize the cooling rate and raise the casting speed in the horizontal-type TRC process, Cu-Cr rolls were used and the castings were conducted in the absence of lubrication on the roll surface. The casting speed was increased to 5 m/min with a decreased RSF. For comparison, strips were also cast at 0.8 m/min using a conventional horizontal caster from a mass production line. Figure 1 shows the inverse pole figure (IPF) maps and pole figures (PFs) of the Al-5Mg strips fabricated at different casting speeds. In the case of the 0.8 m/min condition, the surface region exhibited a shear deformation texture (rot-Cube, {111}//ND), while a typical rolling texture (brass, S, Copper components) was dominant in the central region, thereby indicating that the strip was hot-rolled under high RSF conditions. In contrast, for the 5 m/min condition, the strip exhibited an equiaxed grain structure with an overall random texture. This implies that the rapid solidification structure instead of the hot-rolled grain structure can be obtained at an increased casting speed. It is therefore necessary to consider what role the RSF plays under high casting speeds during the TRC process.

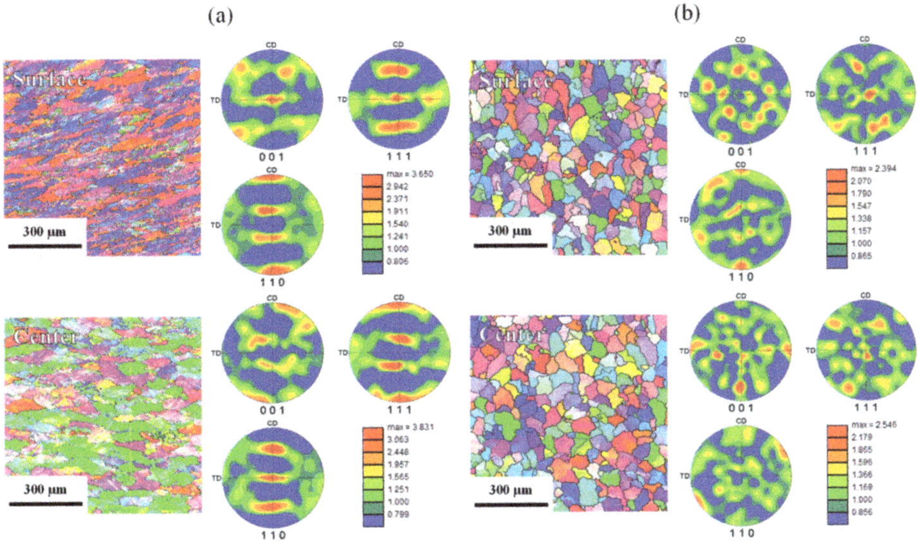

Figure 1. Inverse pole figure (IPF) maps and pole figures (PFs) of twin-roll cast strips fabricated at casting speeds of (**a**) 0.8 m/min, and (**b**) 5 m/min.

3.2. Cooling Behavior of the Strip in HSTRC

Using a vertical-type caster, HSTRC was carried out under various RSF conditions (i.e., 3, 20, and 60 kN) with a casting speed of 60 m/min. The temperature distribution from the center line of the melt pool to the strip center was then investigated using the direct temperature measurement technique. Figure 2 shows the change in temperature in the area of the roll nip. As indicated, the temperature began to decrease significantly at the kiss point where two solidifying shells encountered one another prior to the strip passing through the roll gap. At the roll nip, the cooling rate exhibited its maximum value, and the temperature of the strip center further decreased even after expulsion of the strip from the roll gap, due to the thermal gradient along the thickness direction. This was followed by the air cooling region where the strip temperature remained relatively constant. Using an RSF of 3 kN, the cooling rate at the roll nip was relatively low, resulting in the final strip temperature being equal to the eutectic temperature. As the RSF was increased to 20 kN, the cooling rate increased sharply, although no significant change in the cooling rate was observed upon increasing the RSF further to 60 kN. This result indicates that the contact condition between the strip and the roll surface was enhanced upon increasing the RSF, thereby resulting in the observed increased cooling rate. In the HSTRC process, the cooling behavior of the section from the kiss point to the position of the lowest strip temperature is important, as it determines the final strip temperature. If the cooling rate is decreased, i.e., the contact condition between the strip and the roll surface is poor due to solidification shrinkage, some liquid can remain at the mid-thickness region of the strip. This can cause severe internal cracking or tearing of the strip during the continuous casting process due to the low nature of the high temperature stiffness of aluminum alloys [4,5]. Moreover, if the final strip temperature is higher than or equal to the eutectic temperature, the solidification structure can become coarse under air cooling.

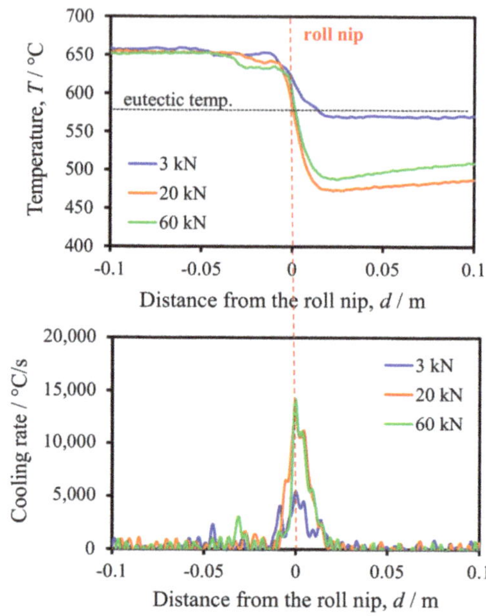

Figure 2. Changes in the (**a**) temperature and (**b**) cooling rate in the mid-thickness region of the strips under various roll separating forces.

Figure 3 shows the microstructures of the Al-7Si-3Mg alloy strips fabricated under RSF conditions of 3 and 20 kN. In the case of the 3 kN condition, relatively coarse plate-like eutectic Si structures were observed, implying that solidification of the strip proceeded under air cooling (i.e., under a low cooling rate) [6]. In contrast, fine dendrite and rod-like eutectic Si was observed in the strip fabricated at 20 kN, indicating that solidification took place during rapid cooling in the area of the roll nip. As such microstructural differences can greatly affect the mechanical properties of the strip [7], the application of an appropriate RSF is therefore critical for the HSTRC fabrication of sound aluminum strips with fine microstructures.

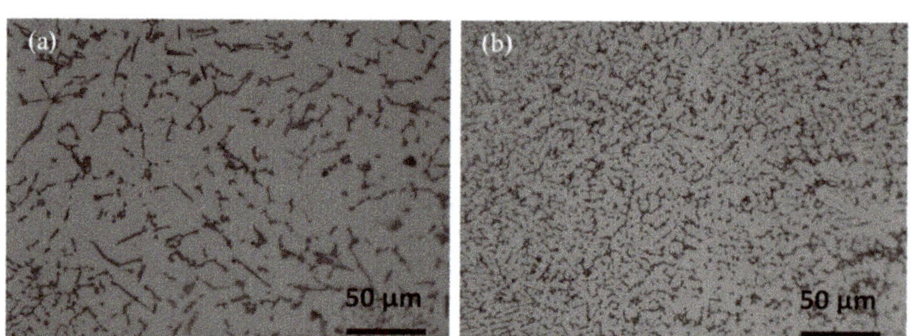

Figure 3. Microstructures of the Al-7Si-0.3Mg alloy strips fabricated using RSF (role separating force) values of (**a**) 3 kN and (**b**) 20 kN.

3.3. Microstructure of the HSTRC Strip

In the HSTRC process, the final solidification region of the strips (generally the mid-thickness region) is greatly affected by the RSF. Especially in the case of aluminum alloys exhibiting a wide range of freezing temperatures, solidification proceeds with the formation of a mushy layer at the growth front of the solidifying shell on the roll surface. Consequently, when two shells encounter one another at the kiss point, the microstructure of the semi-solid mid-thickness region is influenced by the external stress. Thus, Figure 4 shows the macro- and microstructures of the Al-2Si strip fabricated at a casting speed of 60 m/min and an RSF of 20 kN. As shown, an equiaxed grain macrostructure was apparent at the outer shell region, while a band of fine globular grains formed in the mid-thickness region. The formation of this globular grain band is likely related to the compression mode under the different RSF conditions. It is noted that Suery and Flemings conducted a simple compression test on semi-solid dendritic alloys and demonstrated the fragmentation of dendrite arms under a high compression rate [8]. In the HSTRC process, the high casting speed and the RSF conditions caused the active fragmentation of dendrite arms in the semi-solid mushy layer, resulting in the formation of a band structure in the mid-thickness region. This microstructural feature is significant in relation to center segregation control.

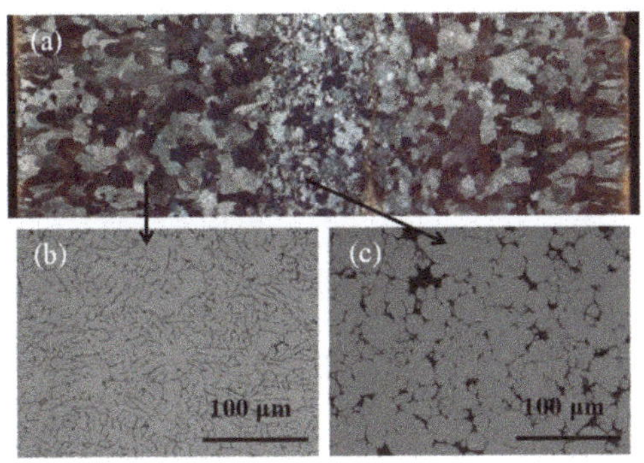

Figure 4. (**a**) Anodized grain structure, and normal etched microstructures of (**b**) the outer shell, and (**c**) the central band region.

3.4. Influence of the Center Segregation Type on the Mechanical Properties of the TRC Strips

In a previous study, it was shown that the center segregation pattern changed with variation in the casting speed [3]. For the horizontal-type TRC of Al-5Mg alloys, typical channel segregation took place under a relatively low casting speed (3 m/min) and a high RSF (Figure 5a). In contrast, the segregation pattern changed to a segregation band consisting of a mixture of an Mg-rich phase and α-Al as casting speed was increased to the high-speed range, i.e., 5 m/min, which is similar to the microstructure shown in Figure 4c (Figure 5b).

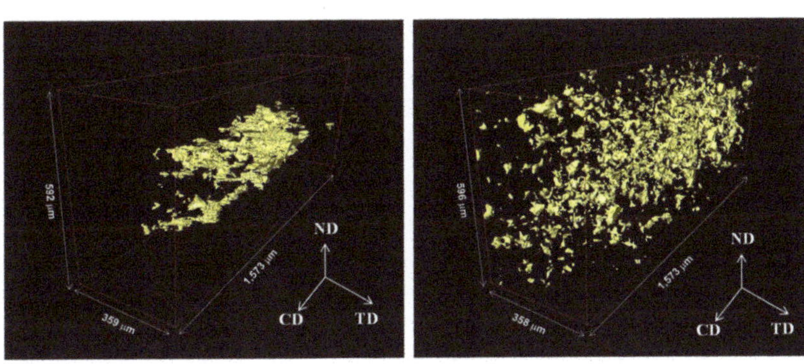

Figure 5. 3D images of center segregation in the strips fabricated at (**a**) 3 m/min, and (**b**) 5 m/min.

To investigate how the center segregation type affects the mechanical properties of the strips, tensile tests were conducted in both the CD and TD. Figure 6 shows the engineering stress-strain curves and SEM images of the resulting fracture surfaces. The SEM images on the right-hand side show the magnified microstructure around the center segregation zones of the left-hand images. The obtained results show that the majority of the outer shell region consisted of dimples, which indicate ductile fracture, whereas the Mg-rich segregation zone revealed cleavages, which suggest brittle fracture. For the 3 m/min conditions, the TD sample exhibited a continuous cleavage along the channel segregation (Figure 4c), causing a premature fracture with a rapid stress drop during the tensile test (Figure 4a). In contrast, the 5 m/min samples showed a region of connected dimples and discrete cleavage in the segregation band, and no drastic premature fracture was observed in both the CD and TD, thereby indicating that formation of this band structure should be beneficial in controlling center segregation in the HSTRC process.

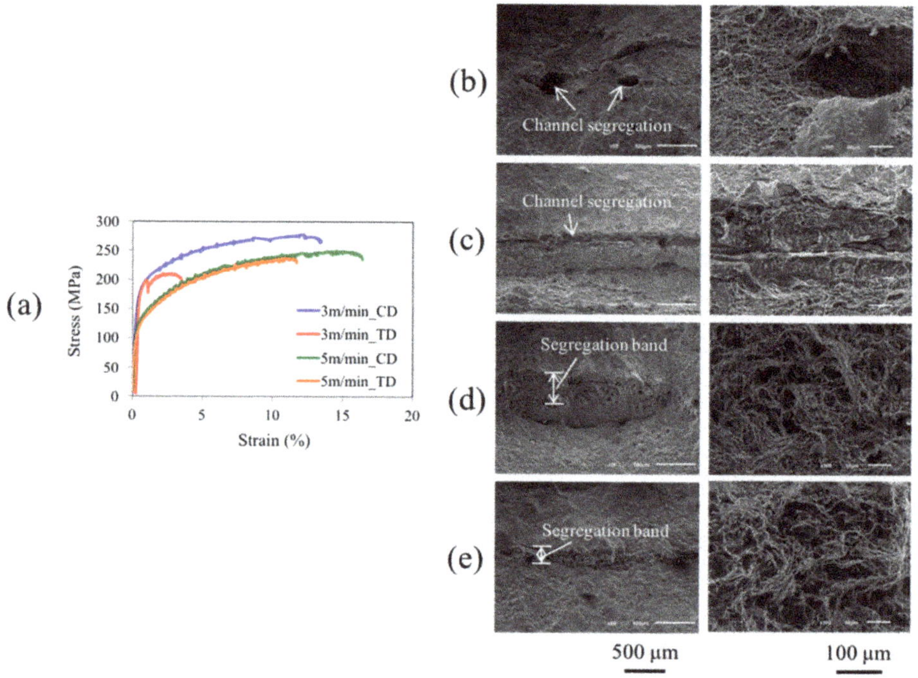

Figure 6. (**a**) Engineering stress-strain curves and SEM images of the fractured surface of the (**b**) 3 m/min, CD, (**c**) 3 m/min, TD, (**d**) 5 m/min, CD, and (**e**) 5 m/min, TD samples.

4. Conclusions

In the traditional twin-roll casting (TRC) process of aluminum alloys, the caster was designed to combine metal casting and hot rolling into a single operation. However, the traditional TRC process is generally operated at low casting speeds (<2 m/min). Thus, to increase the casting speed, the TRC concept was amended from hot rolling to rapid solidification through the application of a minimum roll separating force (RSF). Upon increasing the casting speed with a reduced RSF, it was found that the strip texture changed from a rolling texture to a random texture. In addition, in the case of high-speed TRC (HSTRC), direct temperature measurements indicated that the RSF played an important role in maintaining a good contact between the strip and the roll surface during the casting process. This in turn aided in the fabrication of a sound continuous strip with a fine microstructure. Moreover, the present RSF conditions allowed the formation of a band structure consisting of fine globular grains in the mid-thickness region of the strip, which aids in producing a well-dispersed center segregation pattern. The formation of a segregation band structure was beneficial for the fabrication of a sound strip with no scattering of the mechanical properties.

Author Contributions: Conceptualization, M.-S.K. and S.K.; methodology, M.-S.K. and S.-H.K.; validation, M.-S.K. and S.K.; formal analysis, M.-S.K. and S.-H.K.; investigation, M.-S.K.; resources, M.-S.K.; data curation, M.-S.K.; writing—original draft preparation, M.-S.K.; writing—review and editing, H.-W.K. and S.K.; visualization, M.-S.K.; supervision, H.-W.K. and S.K.; project administration, H.-W.K. and S.K.; funding acquisition, H.-W.K. and S.K.

Funding: This work was supported by the National Strategic Program funded by Ministry of Trade, Industry and Energy (No. 10081329), Republic of Korea.

Conflicts of Interest: The authors declare no conflict of interest.

References

1. Haga, T.; Takahashi, K.; Ikawa, M.; Watari, H. A vertical-type twin roll caster for aluminum alloy strips. *J. Mater. Process. Technol.* **2003**, *140*, 610–615. [CrossRef]
2. Haga, T.; Takahashi, K.; Ikawa, M.; Watari, H. Twin-roll casting of aluminum alloy strips. *J. Mater. Process. Tech.* **2004**, *153–154*, 42–47. [CrossRef]
3. Kim, M.S.; Kim, S.H.; Kim, H.W. Deformation-induced center segregation in twin-roll cast high-Mg Al-Mg strips. *Scr. Mater.* **2018**, *152*, 69–73. [CrossRef]
4. Kim, M.S.; Kim, H.W.; Kumai, S. Direct temperature measurement of Al-2mss%Si alloy strips during high-speed twin-roll casting and its application in determining melt/roll heat transfer coefficient for simulation. *Mater. Trans.* **2017**, *58*, 967–970. [CrossRef]
5. Kim, M.S.; Kumai, S. Effect of Si content on strip thickness and solidified structure in high-speed twin-roll cast Al-Si alloy strips. *Mater. Trans.* **2011**, *52*, 856–861. [CrossRef]
6. Aguilera-Luna, I.; Castro-Román, M.J.; Escobedo-Bocardo, J.C.; García-Pastor, F.A.; Herrera-Trejo, M. Effect of cooling rate and Mg content on the Al-Si eutectic for Al-Si-Cu-Mg alloys. *Mater. Charact.* **2014**, *95*, 211–218. [CrossRef]
7. Jiang, B.; Ji, Z.; Hu, M.; Xu, H.; Xu, S. A novel modifier on eutectic Si and mechanical properties of Al-Si alloy. *Mater. Lett.* **2019**, *239*, 13–16. [CrossRef]
8. Suery, M.; Flemings, M.C. Effect of strain rate on deformation behavior of semi-solid dendritic alloys. *Metall. Trans. A* **1982**, *13*, 1810–1819.

metals

MDPI

Article

Cavity Behavior of Fine-Grained 5A70 Aluminum Alloy during Superplastic Formation

Sheng Li, Shunyao Jin and Zhongguo Huang *

School of Mechanical Engineering, University of Science and Technology Beijing, Beijing 100083, China; lishengsir@163.com (S.L.); jinshunyao@ustb.edu.cn (S.J.)

* Correspondence: jsy_white@126.com; Tel.: +86-010-6233-2467

Received: 18 November 2018; Accepted: 12 December 2018; Published: 14 December 2018

Abstract: The study of the exact physical mechanism of cavity nucleation and growth is significant in terms of predicting the extent of internal damage following superplastic deformation. The 5A70 alloy was processed by cold rolling for 14 passes with a total reduction deformation of 90% (20–2 mm) and the heat treatment was inserted at a thickness of 10 and 5 mm at 340 °C for 30 min. The superplastic tensile tests were performed at 400, 450, 500, 550 °C and the initial strain rate was 1×10^{-3} s^{-1}. Cavities were observed at the head of the particle and the interface of the grain boundaries. It is suggested that the cavity was nucleated during the sliding/climbing of the dislocations, due to the precipitate pinning effect and the impeding grain boundary during grain boundary sliding (GBS). In this study, the results demonstrated a clear transition from diffusion growth to superplastic diffusion growth and plastic-controlled growth at a cavity radius larger than 1.52 and 13.90 μm. The cavity nucleation, growth, interlinkage and coalescence under the applied stress during the superplastic deformation, as well as the crack formation and expansion during the deformation, ultimately led to the superplastic fracture.

Keywords: 5A70 aluminum alloy; superplastic tensile; cavity growth; fracture

1. Introduction

The massive magnesium solid solution from the dispersion phase and the magnesium content in aluminum alloys improve the strength properties, fatigue strength, wear resistance, joint properties and the forming properties [1–3]. Aluminum alloys have been extensively researched with regard to superplastic formation due to their non-heat-treatment properties. In recent decades, aluminum alloy superplastic structures have been cited in the locations of Su-27, MIG-26 and J-10 stamping parts such as the fuselage, stringer, vertical tail skin, etc. [4]. It is well known that aluminum alloys exhibit superplastic fractures due to the presence of the cavity caused by the Mg-rich phase particles. However, most early studies related a less than 5 mass% magnesium content of aluminum alloys with excellent elongation.

Numerous researchers have reported the superplasticity of the Al–Mg series aluminum alloys [5–8], super-strength aluminum alloys (Al–Mg–Cu series) [9,10] and ultra-high-strength aluminum alloys (Al–Zn–Mg–Cu series) using equal channel angular pressing (ECAP), friction stir processing (FSP) and high-pressure rotation (HPT) methods [11–14]. It demonstrates that superplastic flow can be achieved in aluminum alloy with small grain sizes, typically less than 10 μm. In order to satisfy the requirement of a 2 mm thickness of superplastic stamping products, the precipitated phases in the ingot casting billet were controlled, combined with heat treatment between the cold rolling of the 5A70 aluminum alloy. The obtained superplastic elongation-to-failure value, δ, was greatly improved when compared to the existing commercial sheets [15].

In this study, attention was focused on the exact mechanism of cavity nucleation and cavity growth during the superplastic deformation to study the cavity behavior in detail. The cavity behavior found at different tensile stages and in different superplastic fractures were precisely identified via supporting microscopy evidence. However, cavitation is a function of external factors such as temperatures and grain sizes, which are both related to the second phase particles. It was found that the cavity nucleation, growth, interlinkage and coalescence during superplastic deformation were caused extensively by the precipitated phase [16–20].

2. Materials and Methods

The chemical composition of the 5A70 alloy is shown in Table 1. The prepared experimental alloy ingot was homogenized and annealed at 450 °C for 40 h. Then the ingot alloy was rolled into a billet with a rectangular normal direction plane of 255 mm × 255 mm at 380 °C. From the state of the extruding ingot, a billet with a plate shape (200 mm × 200 mm × 25 mm) was processed, and the natural aging treatment was ≥240 h. The size and distribution of the precipitated phases in the smelted and forged processes were controlled to promote the nucleation during dynamic recrystallization, and the billet was subjected to 14 passes of cold rolling on a 350 mm reversing cold mill, and full recrystallization was inserted when the sheet was 10 and 5 mm thick using a GS-2-1200 box-type resistance furnace (Tianjin Zhonghuan Lab Furnace Co., Ltd., Tianjing, China). Ultimately, the 2 mm thickness of fine-grained (FG) 5A70 alloy superplastic sheet was obtained, and the initial grain size of the rolling direction plane was 8.48 μm. Further details of the rolling process and the full recrystallization system were reported in a previous work [15].

Table 1. Chemical compositions of the studied 5A70 aluminum alloy (wt.%).

Mg	Fe	Cu	Si	Zn	Ti	Mn	Al
5.72	0.20	0.058	0.080	0.020	0.043	0.60	Bal.

Specimens with 8 mm gage length and 4 mm gage width were machined in the parallel rolling direction. Superplastic tensile tests were performed on an AG-250KINC Instron machine (Shanghai Gold Casting Instrument Analysis Co., Ltd., Shanghai, China) with a microprocessor control pad in the NV63-CV high temperature furnace. The tests were performed at different temperatures ranged from 400 to 550 °C, and the initial strain rate was 1×10^{-3} s^{-1} in an air condition. In addition, type-K thermocouples were used to detect the furnace and the temperature was controlled within approximately ±2 °C along the entire gage length during the tests. The specimen was insulated at 340 °C and 10 min for full recrystallization; moreover, it can control the temperature equilibrium during the heating process. Immediately, the same heating rate, 21 °C/min, was utilized to reach the target temperature and maintained for 2 min.

The microstructural characterization and analysis of the 5A70 alloy were carried out using a Jeol-7100 (JEOL Ltd., Tokyo, Japan) transmission electron microscope (TEM) and a 600FEG (FEI Corporation, Hillsboro, OR, USA) scanning electron microscope (SEM) or structural characterization analysis. The average grain size was found using the linear intercept method and the OIM software (6.2.0 x86 version, EDAX Inc., Draper, UT, USA). The boundary orientation was measured by utilizing pixel-to-pixel measurements. To carry out the structural characterization, specimens of the SEM were cut 5 mm from the superplastic fracture surface.

3. Results and Discussion

3.1. Superplastic Tensile Tests

It is known that 5A70 aluminum alloy is non-heat treatable with the dissolution and melting temperature is 556 and 631 °C, respectively. Superplastic tensile tests focused on investigating the states of cavity in superplastic deformation of the studied 5A70 alloy. Superplastic tensile tests of FG 5A70 alloy were performed at an initial strain rate of 1×10^{-3} s^{-1} and the temperatures ranged from 400 to 550 °C, according to relevant literature and thermal analysis experimental results [21].

The true stress–true strain results of the superplastic tensile tests are shown in Figure 1a. At a constant strain rate, the strain enhancement phenomenon of the materials was consistent with the general law of superplastic elongation characteristics and increased temperature. Meanwhile, three more tensile tests were carried out and the tensile deformations were unloading at the strain $\varepsilon = 0.65$, 1.40 and 2.65, when the superplastic tensile test temperature was 500 °C and the strain rate was 1×10^{-3} s^{-1}. The intermediate state of the deformation structures was obtained using water quenching, and the engineering stress–engineering strain results are shown in Figure 1b.

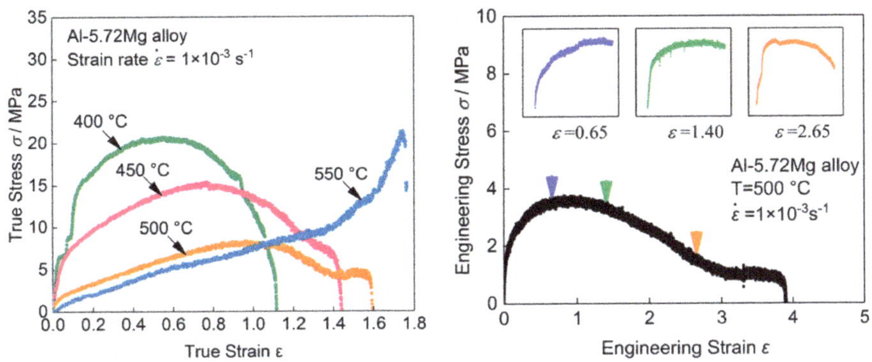

(**a**) (**b**)

Figure 1. True stress–true strain results of 5A70 aluminum alloy at 400–550 °C and 1×10^{-3} s^{-1} (**a**), and different superplastic tensile strain stages, $\varepsilon = 0.65$, 1.40 and 2.65, at 500 °C and 1×10^{-3} s^{-1} (**b**).

Figure 1a shows that there was no obvious steady-state flow stage in the superplastic tensile state of the 5A70 alloy, as was the case for other aluminum alloys with a different magnesium content [22,23]. When the strain rate was constant, the superplastic elongation-to-failure results of the FG 5A70 alloy enlarged with the increased temperatures were δ = 205%, 321%, 390% and 406%. At 550 °C, which is close to the dissolution temperature of precipitated phase particles. The reduction of the pinning effect during dynamic recrystallization because of the content of phase particles decrease. In addition, the strain hardening obviously occurred during the superplastic deformation due to the increased distortion of the grain growth, leading to an increase of true stress [24]. The strain hardening was attributed to dislocation sliding/climbing. Additionally, the dislocation density changed nonmonotonically with stable grain structure during the initial stage of the superplastic deformation. A high dislocation density at the beginning of the deformation at T = 400 °C with a high strain rate caused by grain adaptation—i.e., where the dislocation density increased rapidly and was plugged into the grain—formed dislocation walls/cells and led to increased true stress [25]. However, with the accumulation of deformations, the grain rotation occurred under shear stress and dislocations were absorbed by the grain boundaries, which led to the true stress remaining stable for a short period of time [26]. This is the reason for the true stress presenting in a step-up state. With the increasing temperature and the accumulation of superplastic deformation, the true stress increased and the strain hardening strengthened gradually and clearly.

Figure 1b demonstrates that in the initial stage of superplastic tensile, the engineering stress increases continuously and reaches its maximum at $\varepsilon = 0.65$. This is due to the large number of dislocations formed during superplastic deformation and the grain growth in high temperature and low strain rate. The maximum stress value, σ_{max} = 3.75 MPa as shown in Figure 1b. When the strain reached $\varepsilon = 1.40$, the applied stress gradually decreased and the superplastic deformation reached a correspondingly stable region. The applied stress slowly decreased and eventually remained low value until the strain reached $\varepsilon = 2.65$. The data at different stages of superplastic tensile stress are provided by the test results in Figure 1b. In the last stage of the superplastic tensile, the stress value remains constant in Figure 1b and the corresponding true stress value increases continuously as shown in Figure 1a.

3.2. Cavity Nucleation

The superplastic tensile specimen was analyzed using a TEM at 500 °C and 1×10^{-3} s^{-1} with the strain ε = 0.65, as shown in Figure 2. This illustrates that the increase of stress at the initial stage of the superplastic tensile stage due to the increase of the dislocation density under the grain boundary sliding (GBS), in addition, the density of the dislocation was ~3.65×10^{14} m^2.

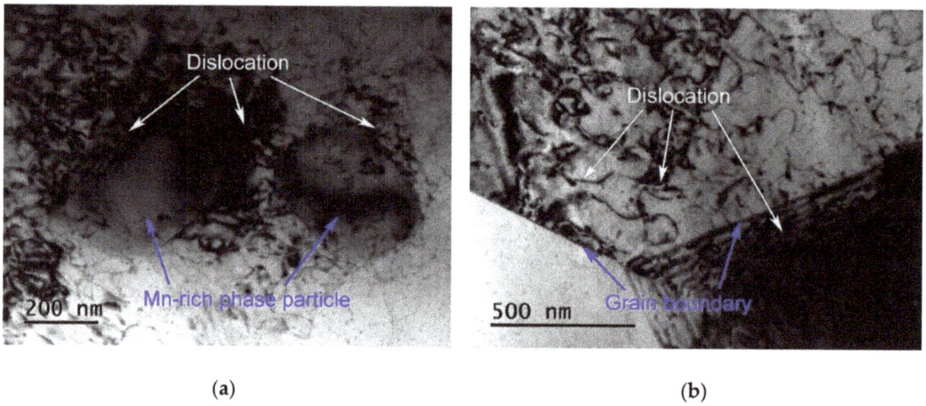

Figure 2. The relationship between the dislocations, precipitate phases (**a**) and grain boundaries (**b**) found using the TEM tests at 500 °C and 1×10^{-3} s^{-1} (ε = 0.65).

The diffusion activation energy ranged from 135 to 139 kJ/mol at temperatures ranged from 400 to 500 °C, which were close to the lattice diffusion activation energy, 143.4 kJ/mol, of pure aluminum [27]. Analysis of the superplastic behavior in terms of the diffusion activation energy and surface morphology illustrates that lattice diffusion dominates the GBS deformation mechanism of FG 5A70 alloy, and the GBS occurs through the dislocation sliding/climbing on grain boundaries. At 500 °C and 1×10^{-3} s^{-1}, when the applied stress accumulation reached the maximum point in superplastic tensile stage, it was clearly found that under the shear stress the GBS caused dislocations to pile up near the precipitation phase, as shown in Figure 2a, and dislocations crossed the grain boundary by sliding and climbing to form a subgrain boundary, as shown in Figure 2b. The dislocation density was higher and the plugging/gathering occurred at the head of the precipitation phases, leading to the stress concentration. When the pile-up stress, σ_p, exceeded the theoretical decohesion strength of the (Al-matrix/Second phase particles) interphase boundary, a small cavity formed, and the cavity began to nucleate attached to the precipitation phase [28]. The stress at the head of the pile-up, σ_p, is given by [29]:

$$\sigma_p = \frac{2L\tau^2}{Gb}, \tag{1}$$

where L is the length of the pile-up and equivalent to the linear intercept grain size. G is the shear modulus (MPa). The pure aluminum temperature control model was adopted. In this paper, $L = d/1.74$, $\tau = \sigma/\sqrt{3}$. b is the Burgers vector, $b = a/\sqrt{2} = 2.863 \times 10^{-10}$ m [30]. When the test temperature was 500 °C, the maximum applied stress (σ_{max} = 3.75 MPa) solution was plugged into the type product stress σ_p = 7.87 MPa. The plugging stress threshold was more than twice the applied stress in the superplastic tensile state. It is obvious that the Al matrix and the strengthening phase particles were easily separated, which promoted the cavity nucleation under the different stress levels. In addition, dislocations coalesced at the grain boundary to form a subgrain boundary, as showed in Figure 2b. It is assumed that the shape of the cavity is circular and the equivalent radius of the cavity is r. The change of the Helmholtz free energy of the system was obtained as follows [27,31,32]:

$$\Delta G = -2.53(\frac{2}{3} \times \frac{d}{1.74b} \times \frac{\sigma^2}{G})r^3 + (9.31\gamma - 0.5 \times 2.93\gamma) - 3.79 \times \frac{(\frac{2}{3} \times \frac{d}{1.74b} \times \frac{\sigma^2}{G})^2}{2E} r^3, \quad (2)$$

where γ is the surface energy of the cavity, d is the grain size and E is the Young's modulus of pure aluminum. Figure 3 shows the change of the Helmholtz free energy as a function of the cavity radius of the 5A70 alloy deformed at different tensile temperatures, with the strain $\varepsilon = 0.3$ and the strain rate was $\dot{\varepsilon} = 1 \times 10^{-3}$ s^{-1}. By increasing the tensile temperature range from 400 to 550 °C, the maximum values of the Helmholtz free energy ranged from 8.32×10^{-16} to 8.41×10^{-14} J, and the corresponding critical radius of the cavity nucleation maximum values increased from 2.16×10^{-8} to 2.37×10^{-7} m, indicating that the cavity was very different and nucleated at a higher superplastic tensile temperature. That is, at 500–550 °C, it was more difficult for the cavity to cross the nucleation barrier than at 400–450 °C. According to Figure 1a, it is suggested that when the superplastic tensile temperatures were 500 and 550 °C, the strain hardening led to an increase of the true stress. Meanwhile, the cavity nucleation was beneficial to the superplastic flow.

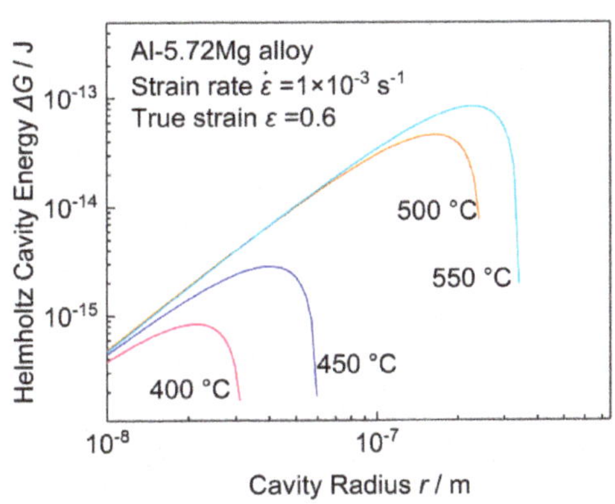

Figure 3. The relationship between the nucleation free energy and cavity radius of 5A70 aluminum alloy in the superplastic tensile state (400–550 °C).

3.3. Cavity Growth

3.3.1. Cavity Growth Controlled by Diffusion and Plastic

The mechanism of cavity growth in the superplastic tensile state is divided into two types: (1) the growth mechanism of stress promoting a spherical cavity growing by diffusion along the grain boundary; and (2) the cavity growth is controlled by the plastic flow of the surrounding material [27,32–35].

At 500 °C and 1×10^{-3} s^{-1}, the parallel sections of the superplastic tensile specimens were tested by the SEM with an aborted strain of $\varepsilon = 0.65$, 1.40 and 2.65; the corresponding results are presented in Figure 4. There are a large number of finely-dispersed second phase particles in the 2 mm thick deformed sheet, and the black points/areas show the cavity distribution during the superplastic tensile deformation. It is clear that the number and density of the cavities increased immediately, accompanied by the constant accumulation of the deformation in Figure 4a,c,e. In addition, the corresponding magnification is shown in Figure 4b,d,f.

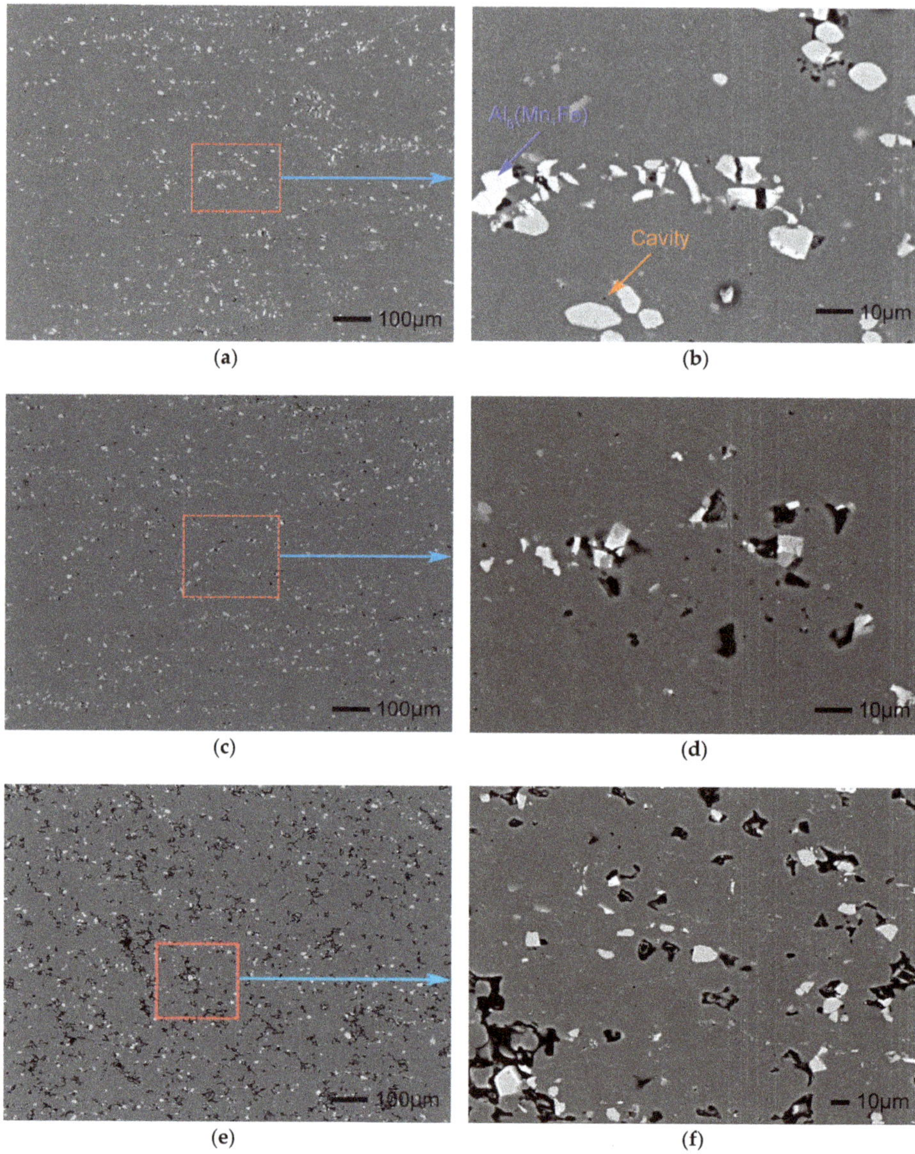

Figure 4. The scanning electron microscopy results in the superplastic tensile stages and magnified regions at 500 °C and 1×10^{-3} s^{-1}: ε = 0.65 (**a**,**b**), ε = 1.40 (**c**,**d**) and ε = 2.65 (**e**,**f**).

Figure 4b,d shows the cavity growth with a strain increase from ε = 0.65 to 1.40, and the cavities evidently grew and interlinked at ε = 2.65 in Figure 4e,f. The initial nucleation stage of the cavity mostly exhibited an O-shape due to the low surface energy of the spherical cavity, which remained relatively stable and easy to nucleate. O-shaped cavities were found at every stage, indicating that new cavity nucleation continued to occur during the process of strain accumulation in the superplastic tensile stage, as shown in Figure 4b,d,f.

At 500 °C and 1×10^{-3} s^{-1}, it is suggested that the applied stress reached the maximum value at ε = 0.65, as shown in Figure 1b. This is because the dislocations continuously accumulated in the second phase particles and the grain boundaries during deformation, as shown in Figure 2, which increased the strength of the 5A70 alloy and provided the driving force for cavity nucleation. Meanwhile, Mg atoms precipitated from the Al-matrix to form second phase particles. The pinning effect of the precipitated phase was encouraged to enhance the cavity nucleation and the chemical potential between the forceful grain boundary atoms and the free surface of the cavity [34,35]. However, the cavity growth rate under diffusion control was obtained as follows:

$$\frac{dr}{d\varepsilon} = \alpha_0\left(\frac{2\Omega\delta D_{gb}}{r^2kT}\right)\left[\frac{\sigma - (2\gamma/r)}{\dot{\varepsilon}}\right], \tag{3}$$

where Ω is the atomic volume, δ is the grain boundary width, $\delta = 2b$. D_{gb} is the coefficient for grain boundary diffusion, k is Boltzmann's constant, T is the absolute temperature, σ is the applied stress, γ is the surface energy, r is the cavity radius, α_0 is the cavity spacing, $\dot{\varepsilon}$ is the strain rate and α_0 is defined as $\alpha_0 = 1/\{4\ln(a/2r) - [1-(2r/a)^2][3-(2r/a_0)^2]\}$. The cavity size parameter was $\alpha_0 = 0.12$ as the initial stage of cavity diffusion complied with $a/2r \geq 10$.

The cavities grew due to plastic deformation in the surrounding crystalline lattice. The plastic-controlled growth mechanism is given by [33–35]:

$$\frac{dr}{d\varepsilon} = r - \frac{3\gamma}{2\sigma}, \tag{4}$$

where the superplastic tensile of 5A70 aluminum alloy was tested at 400–500 °C, and the critical cavity radius, r_c, denoting the transition from conventional diffusion growth for plastic-controlled growth was obtained from Formulas (3) and (4), so that [27,33]:

$$r_c = \left(\frac{2\Omega\delta D_{gb}}{kT}\right)^{\frac{1}{3}}\left(\frac{\sigma}{\dot{\varepsilon}}\right)^{\frac{1}{3}}, \tag{5}$$

Formula (5) demonstrates that when the temperature and strain rate are constant, the lower the applied stress in the superplastic tensile deformation, and the smaller the critical cavity radius, r_c, exhibited. Figure 4 shows the tension strain ε = 0.65, 1.40 and 2.65, presenting the results of the SEM analysis as pointed in the Figure 1b. As noted above, the accumulation of tensile deformation causes a decrease in the critical nucleation radius, moreover, the cavity density increases immediately. This is a complex function of cavity nucleation related to the testing temperature, strain rate, microstructure size and phase-particle shape and size. Generally, larger particles and larger interfacial defects cause a more rapid separation of the Al-matrix and second phase particles with an increase in tensile deformation, while smaller particles have a smaller restricted plastic zone, and greater accumulated strain is needed to produce complete separation for continuous cavity nucleation [36,37]. Taking the data in Table 2 into Formulas (3) and (4), the cavity growth rate, *dr/dε*, can be drawn as shown in Figure 5.

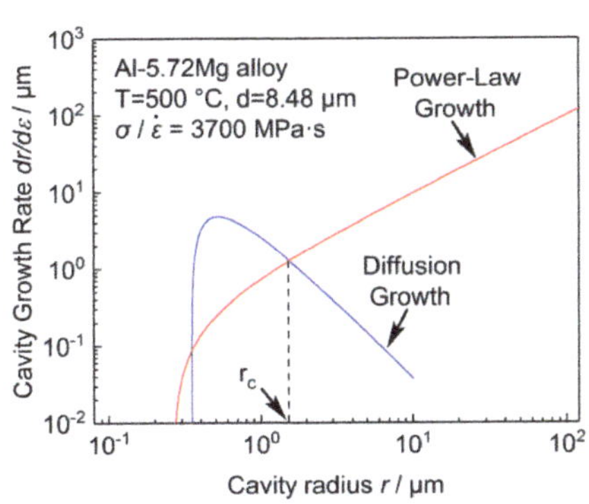

Figure 5. Schematic illustration of diffusion growth and plastic-controlled growth, showing the critical radius r_c.

Table 2. Constants used for the calculation of the cavity growth rate ($dr/d\varepsilon$).

Temperature T/°C	Applied Stress σ/MPa	Strain-Rate Sensitivity m	Atomic Volume V/m^3	Strain Rate ε/s^{-1}
500	3.70	0.47	2.34×10^{-2}	1×10^{-3}

* The material constants used for 5A70 alloy: $\delta = 5.72 \times 10^{-10}$ m, $k = 1.38 \times 10^{-23}$ J·K^{-1}, $D_{gb} = 3.3 \times 10^{-14}$ m^3·s^{-1}, $\gamma = 0.64$ J·m^{-2} and $R = 8.314$ J·mol^{-1}·K^{-1}.

Figure 5 shows the relationship between the cavity growth rate, $dr/d\varepsilon$, and the cavity radius, r, indicating the mechanism of cavity growth during the superplastic tensile state of FG 5A70 alloy. The figure illustrates that when the cavity radius $r < r_c = 1.52$ μm, the cavity growth controlled by diffusion dominates the cavity growth mechanism. By contrast, the cavity growth is controlled by the plastic flow, exhibiting a power-law growth rate that is larger than the other type of growth rate. In addition, during plastic-controlled cavity growth the cavity volume fraction should increase exponentially with strain, and η is the growth rate parameter for the cavity volume [38,39].

$$\eta = \frac{3}{2}\frac{m+1}{m}\sinh\frac{2(2-m)}{3(2+m)} \tag{6}$$

The coefficient of strain rate sensitivity, m, value continuously decreases along with the accumulation of tensile deformation. However, the cavity growth rate parameter increases as the strain increases simultaneously. Figure 6 shows the true strain range from $\varepsilon = 0.1$ to 1.6, while the growth rate parameters increased correspondingly, from $\eta = 1.21$ to 3.22. Therefore, the diffusion-controlled cavity growth with a small growth rate parameter occurred in the initial stage of superplastic tensile deformation. As the plastic-controlled growth dominated the cavity growth, the cavity growth rate parameter and the cavity growth rate increased simultaneously. Meanwhile, it was clearly found that plastic-controlled growth of the cavities occurred mainly due to cavity interlinkage and coalescence.

3.3.2. Effect of Superplastic Diffusion on Cavity Growth

Along with superplastic tensile deformation, the Mg atoms continuously separated from the distorted Al-matrix lattice into the generated vacancies. When the cavity size of the diffusion mechanism was larger than the grain size, the newly formed vacancies diffused into the neighboring cavity along the multiple grain boundaries under stress. Meanwhile, superplastic diffusion growth of the cavity nucleation region appeared, and then the cavities begin to interlink during the growth process. Superplastic diffusion growth was the dominant mechanism during this stage, when the voids vacated into the adjacent cavities, subject to diffusion-controlled growth.

Chokshi et al. [35] noted that the growth of cavities due to superplastic diffusion occurred while the cavity was large enough to intersect with multiple grain boundaries. This required a cavity size larger than half of the grain size, $d/2$, therefore, the grain size of cavity growth under superplastic diffusion was considered between $d/2$ and d.

Figure 6 shows comprehensively the cavity growth parameter, η, cavity area fraction, S_c, and the grain diameter radius, d, with the cavity equivalent radius, r. Meanwhile, the corresponding S_c was presumed to be 0.24, 0.77 and 3.96%, as shown in Figure 4a,c,e.

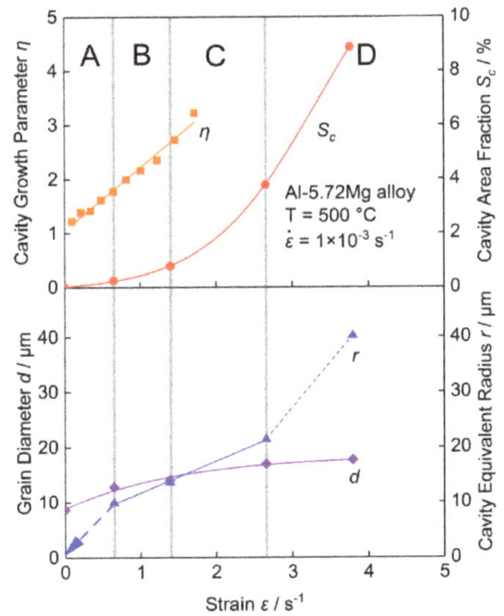

Figure 6. The cavity growth parameter, η, cavity area fraction, S_c, cavity equivalent radius, r, and dynamic recrystallization grain size, d, at different superplastic tensile stages at 500 °C and 1×10^{-3} s^{-1}.

Region A in Figure 6 shows an increase in the applied stress due to increased dislocation and the continuous nucleating of the cavity (Figure 4) at the tensile strain stage ε = 0–0.65, corresponding to cavity equivalent radius r = 9.76 μm and grain size d = 12.83 μm. The dominant mechanism of cavity growth transformed from diffusion growth to superplastic diffusion growth, judging by the inequality $r_c \leq d/2 \leq r \leq d$. In region B, with a tensile strain of ε = 0.65–1.40, the applied stress continued to decline due to the cavity nucleation and cavity growth. Superplastic diffusion growth took effect because the cavity radius complied with $d/2 \leq \mathrm{r} \approx d$, while the cavity equivalent radius was r = 13.90 μm and the grain size was d = 13.96 μm at tensile strain ε = 1.40. Figure 6 shows the cavity growth parameter, η, corresponding to the coefficients of strain rate sensitivity, m, which are distributed in the region A and region B. Therefore, superplastic diffusion growth dominated the cavity growth of tensile strain ε = 0.46–1.40. Ultimately, the plastic-controlled growth dominated the cavity growth, interlinkage and coalescence, when the superplastic tensile strain was ε > 1.40, as shown in region C and D.

Figure 6 demonstrates that dynamic recrystallization occurred during superplastic tensile deformation. Due to the pinning effect of a large number of dispersed Mg-rich phase particles, the grain size was 8.6 μm after rolling and 17.67 μm after superplastic fracture in the ND plane. The cavity size increased continuously related to the superplastic diffusion growth. Previous analysis in Figure 4b,d indicated that the cavity nucleation and growth was not interlinked extensively. However, the cavities interlinked and coalesced rapidly due to the cavity equivalent radius, r, being larger than the grain size, d, which coincided with the previous analysis in Figure 4f.

3.3.3. Effect of the Fine-Grained Structure on Cavity Growth

The above analyses showed that the fine-grained structure played an important part in the superplastic diffusion growth and plastic-controlled growth. Figure 7a showed that fully dynamic recrystallization occurred during the superplastic tensile deformation and the grain size increased, while the result for ε = 2.65 illustrated that the grain size fraction of d = 20–30 μm was clearly much higher than ε = 0.65 and 1.40. It is well-know that the glide plane of the face-centered cubic crystal of FG 5A70 alloy is {1 1 1}, and the grain boundary slip direction is <1 1 0> [40]. Meanwhile, in order to satisfy the balance of grain-to-grain deformation and the balance of the reaction stress in the superplastic tensile state, the GBS occurred mainly in the sliding direction. This clearly clarified the grain rotation and the GBS in the superplastic tensile state of the FG 5A70 alloy, as illustrated in Figure 7b [41].

Figure 7b shows that the fraction of the grain boundary angle at 3.58° was 20.3% (ε = 2.65), which was greater than 5.41% (ε = 0.65) and 3.83% (ε = 1.40). Plastic-controlled growth dominated the cavity interlinkage during deformation because the fraction of large cavities increased obviously from ε = 0.65 to 2.65, as shown in Figure 7c. At ε = 2.65, the cavity area fraction of 20–200 μm^2 was significantly higher than ε = 0.65 and 1.40. This demonstrated that the cavity coalescence formed a larger cavity, which was detrimental to the superplastic tensile state and eventually led to cracking.

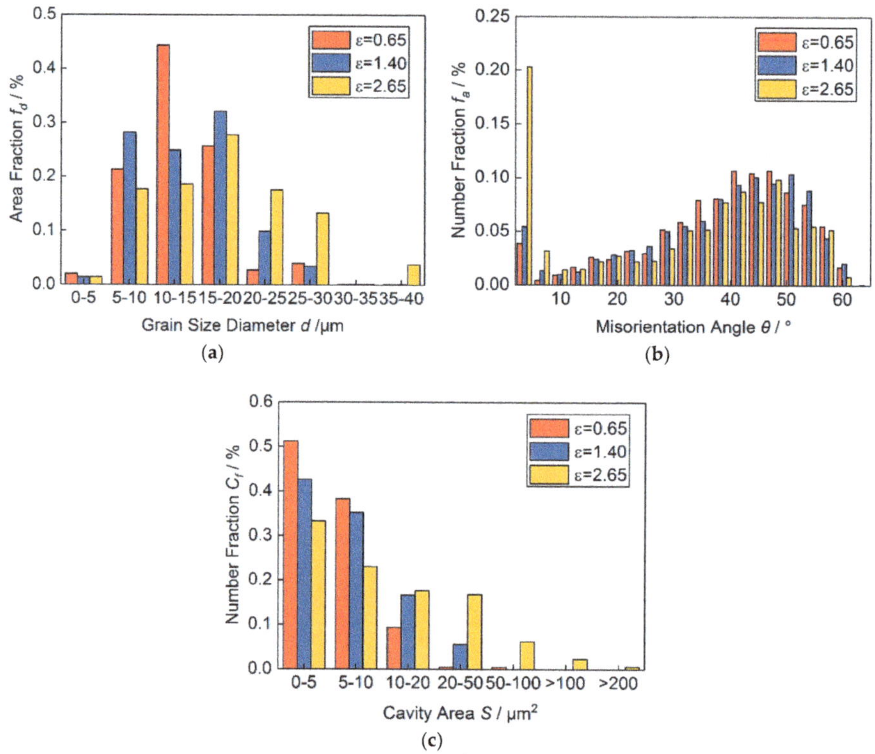

Figure 7. Distribution of the strain due to grain boundary sliding in (**a**) the dynamic recrystallization grain size, (**b**) the grain boundary angle rotation and (**c**) the number of cavity fraction in different strain stages. The superplastic tensile condition was 500 °C and 1×10^{-3} s^{-1} of fine-grained 5A70 alloy.

3.4. Crack Formation

In the present investigation, detailed data were obtained on the formation and development of cavity behavior, suggesting that plastic-controlled growth dominated the cavity interlinkage and coalescence. This was expected to result in crack formation and to influence the superplastic elongation-to-failure of the studied 5A70 alloy.

Figure 8a–h illustrates the testing results of the EBSD close to the fracture locations with different temperatures ranging from 400 to 550 °C, and the strain rate was 1×10^{-3} s^{-1}. Correspondingly, Figure 8i–l shows the morphology tested by the SEM at 5 mm from the superplastic fracture location.

Figure 8 shows the EBSD analyses of the 5A70 alloy deformed at different temperatures at $\dot{\varepsilon} = 1 \times 10^{-3}$ s^{-1} after superplastic fracture. The color of each grain was coded by its crystal orientation based on the [001] inverse pole figure, as seen in Figure 8a. There was an aggregation of a large number of ultrafine grains near the small cavities, in addition, the cavity interlinkage and coalescence were precisely identified via supporting microscopy evidence. Nevertheless, new ultrafine grains occurred in limited regions and generated near initial deformed grains, indicating that dynamic recrystallization had occurred. Figure 8a,c,f,g shows that the microstructure consisted mainly of a grain size larger than 10 μm. It can be clearly seen in Figure 8a,c,e,g that the grain size of the FG 5A70 alloy increased with the increased temperature and the superplastic tensile deformation, while the grain sizes were 9.60, 11.78, 13.32 and 21.16 μm. At 400 °C and 1×10^{-3} s^{-1}, dynamic recrystallization occurred without obvious grain growth. However, in this work the final grain structure had an average recrystallized grain size of less than 15 μm at 400–500 °C, revealing that the 5A70 alloy had a strong ability to inhibit grain growth during superplastic deformation. At 550 °C, the content of Mg-rich phase particles decreased and the abnormal grain growth could be inhibited due to the reduction of the pinning effect during dynamic recrystallization. Therefore, the plastic deformation of the grain growth in the superplastic tensile direction dominated the grain growth and led to crack formation caused by cavity coalescence, as shown in Figure 8g.

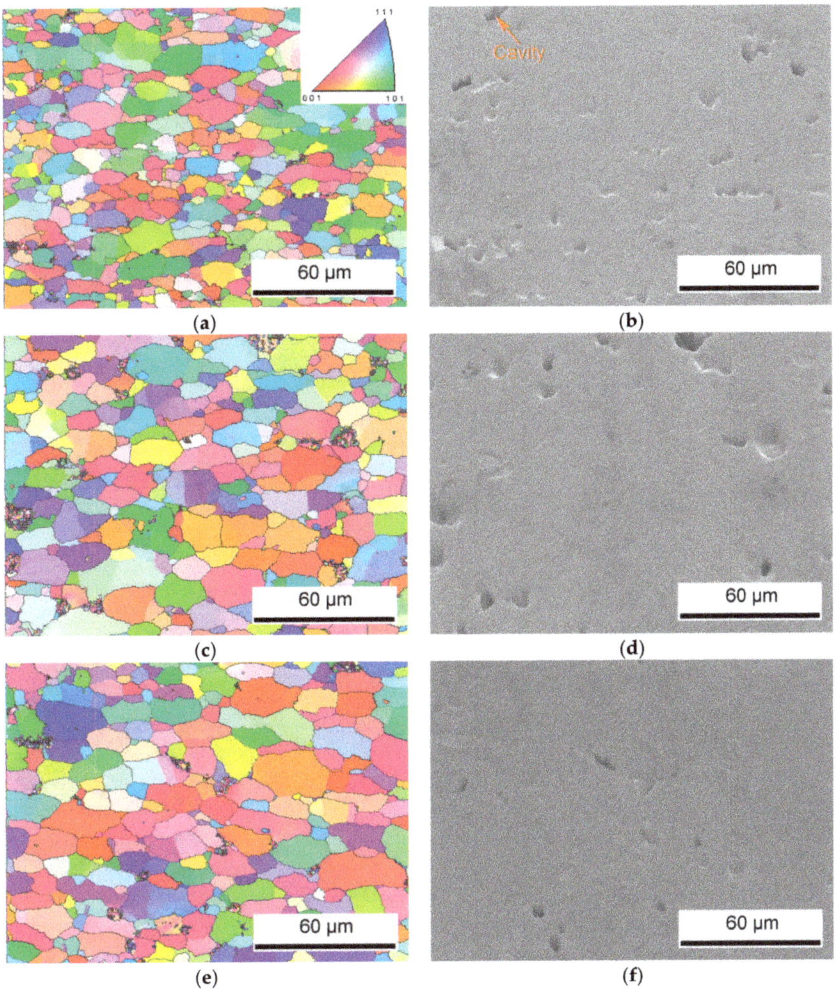

Figure 8. *Cont.*

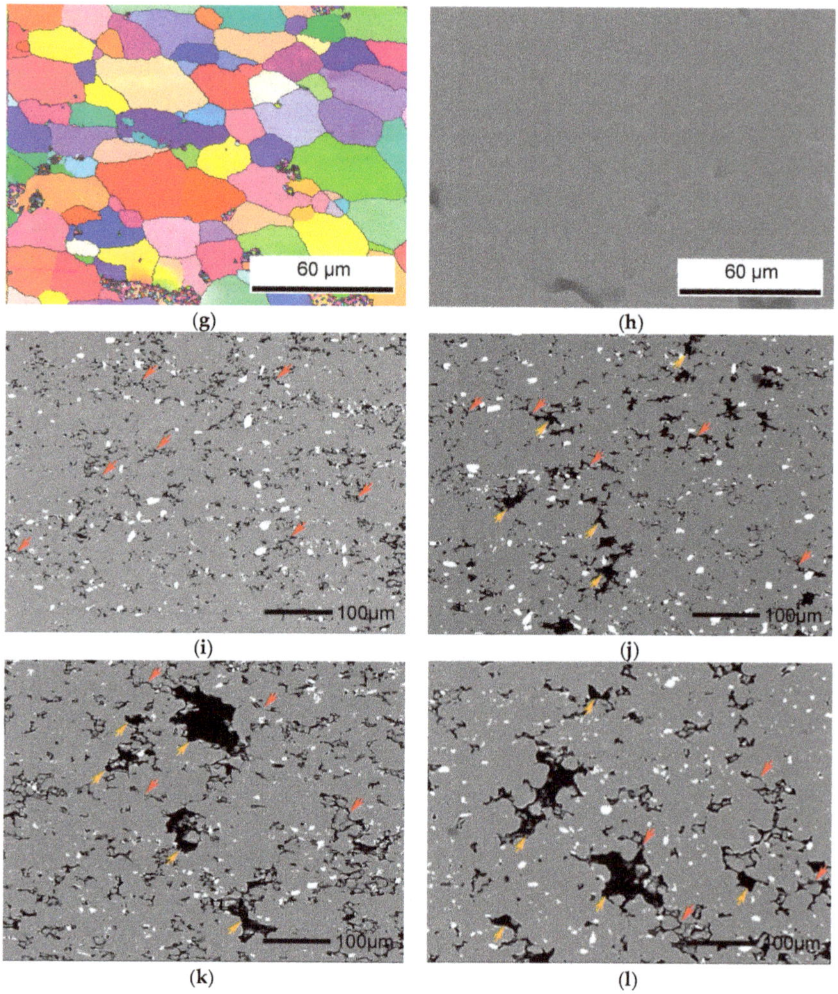

Figure 8. The Grain structures and the superplastic fracture morphologies at initial strain rate 1×10^{-3} s^{-1} with different temperatures: 400 °C (**a**,**b**,**i**), 450 °C (**c**,**d**,**j**), 500 °C (**e**,**f**,**k**) and 550 °C (**g**,**h**,**l**).

In Figure 8i–l it can be seen that the cavities nucleated tightly attached to the particles, the cavity growth interlinked in the superplastic tensile state, indicated by red arrows, and the cavities coalesced, indicated by the yellow arrows. Comparing Figure 4a,c,e and Figure 8a,c,e, it is suggested that with the continuous precipitation of Mg-rich phase particles at the grain boundaries, the new cavities nucleated continuously in the superplastic tensile state. Meanwhile, the cavity area fractions in Figure 8i–k of 1.79, 4.07 and 8.66% increased with temperatures of 400, 450 and 500 °C, respectively. There was a large amount of cavity nucleation and growth along with the accumulation of tensile deformations, as shown in Figure 8i–l. Due to the decrease of second phase particles, it was clearly found that the increase in the grain sizes had a certain effect in terms of promoting the absorption of small cavities at 550 °C and 1×10^{-3} s^{-1}, while the cavity area fraction was 5.66% as shown in Fgiure 8l. Based on this research, it can be stated that cavity nucleation, growth, interlinkage and coalescence, along with the accumulation of tensile deformation, cause crack formation, resulting in the superplastic fracture of FG 5A70 alloys.

4. Conclusions

The following conclusions can be drawn from this work:

(1) Strain hardening occurs in fine-grained 5A70 alloys in the superplastic tensile state at temperatures ranging from 400 to 550 °C, when the strain rate is 1×10^{-3} s^{-1}.
(2) The dislocation density of the fine-grained 5A70 alloy was ~10^{14} m^2, which piled up at the head of the second phase particles and the grain boundaries, resulting in the increase of true stress and the nucleation of the cavities.
(3) Cavity growth mainly occurred at the stage of diffusion growth and superplastic diffusion growth due to the diffusion of the voids into adjacent cavities with a cavity radius smaller than the grain size.
(4) Plastic-controlled growth dominated the cavity interlinkage and coalescence process, which eventually led to superplastic fractures.
(5) At 500 and 550 °C, the abnormal grain growth of FG 5A70 alloy during dynamic recrystallization was higher than 400 and 450 °C due to the decrease in Mg-rich phase particles and the Helmholtz free energies.

Author Contributions: Conceptualization, Z.H. and S.L.; methodology, S.L.; software, S.L.; validation, Z.H., S.J. and S.L.; formal analysis, S.L.; investigation, S.L.; resources, Z.H. and S.J.; data curation, S.L.; writing—original draft preparation, S.L.; writing—review and editing, S.L.; supervision, H.Z.; project administration, S.J.; funding acquisition, S.J.

Funding: This research is funded by the Fundamental Research Funds for the Central Universities of China, Grant number FRF-TP-16-043Al and the APC was funded by S.J.

Acknowledgments: We would like to thank Guoliang Xie for his help in the material preparation, Xuetong Zhao for his assistance with the superplastic tensile tests, Shuming Wang for his assistance with the SEM/EBSD analysis, and Xinqin Liu for her help with the analysis of the results.

Conflicts of Interest: The authors declare no conflict of interest.

References

1. Bolt, P.J.; Lamboo, N.A.P.M.; Rozier, P.J.C.M. Feasibility of warm drawing of aluminium products. *J. Mater. Process. Technol.* **2001**, *115*, 118–121. [CrossRef]
2. Naka, T.; Torikai, G.; Hino, R.; Yoshida, F. The effects of temperature and forming speed on the forming limit diagram for type 5083 aluminum–magnesium alloy sheet. *J. Mater. Process. Technol.* **2001**, *113*, 648–653. [CrossRef]
3. Tokuda, M.; Inaba, T.; Ohigashi, H.; Kurakake, A. Discussions on constitutive equations of superplastic 5083 aluminum alloy. *Int. J. Mech. Sci.* **2001**, *43*, 2035–2046. [CrossRef]
4. Beijing Institute of Aeronautical Materials Group. *Materials Technology of Aeronautics*; Aviation Industry Press: Beijing, China, 2014; pp. 131–143. ISBN 978-7-5165-0298-3.
5. Smolej, A.; Klobčar, D.; Skaza, B.; Nagode, A.; Slaček, E.; Dragojević, V.; Smolej, S. Superplasticity of the rolled and friction stir processed Al–4.5 Mg–0.35Sc–0.15Zr alloy. *Mater. Sci. Eng. A* **2014**, *590*, 239–245. [CrossRef]
6. Charit, I.; Mishra, R.S. Evaluation of microstructure and superplasticity in friction stir processed 5083 Al alloy. *J. Mater. Res.* **2004**, *19*, 3329–3342. [CrossRef]
7. Lee, Y.B.; Shin, D.H.; Park, K.T.; Nam, W.J. Effect of annealing temperature on microstructures and mechanical properties of a 5083 Al alloy deformed at cryogenic temperature. *Scr. Mater.* **2004**, *51*, 355–359. [CrossRef]
8. Verma, R.; Ghosh, A.K.; Kim, S.; Kim, C. Grain refinement and superplasticity in 5083 Al. *Mater. Sci. Eng. A* **1995**, *191*, 143–150. [CrossRef]
9. Soer, W.A.; Chezan, A.R.; Hosson, J.T.M.D. Deformation and reconstruction mechanisms in coarse-grained superplastic Al–Mg alloys. *Acta Mater.* **2006**, *54*, 3827–3833. [CrossRef]
10. Hasegawa, T.; Yasuno, T.; Nagai, T.; Takahashi, T. Origin of superplastic elongation in aluminum alloys produced by mechanical milling. *Acta Mater.* **1986**, *46*, 6001–6007. [CrossRef]

11. Orozco-Caballero, A.; Ruano, O.A.; Rauch, E.F.; Carreño, F. Severe friction stir processing of an Al-Zn-Mg-Cu alloy: Misorientation and its influence on superplasticity. *Mater. Des.* **2018**, *137*, 128–139. [CrossRef]
12. Patel, V.V.; Badheka, V.; Kumar, A. Effect of polygonal pin profiles on friction stir processed superplasticity of AA7075 alloy. *J. Mater. Process. Technol.* **2017**, *240*, 68–76. [CrossRef]
13. Wang, Y.; Mishra, R.S. Finite element simulation of selective superplastic forming of friction stir processed 7075 Al alloy. *Mater. Sci. Eng. A* **2007**, *463*, 245–248. [CrossRef]
14. Garcia-Infanta, J.M.; Zhilyaev, A.P.; Sharafutdinov, A.; Ruano, O.A.; Carreño, F. An evidence of high strain rate superplasticity at intermediate homologous temperatures in an Al–Zn–Mg–Cu alloy processed by high-pressure torsion. *J. Alloy. Compd.* **2009**, *473*, 163–166. [CrossRef]
15. Li, S.; Huang, Z.G.; Jin, S.Y.; Wang, B.Y.; Liu, X.Q.; Lei, K. Heat Treatment Process of Cold Rolled 5A70 Aluminum alloy Superplastic Sheet. *Chin. J. Rare. Met.* **2018**, *42*, 283–288. [CrossRef]
16. Chokshi, A.H.; Langdon, T.G. The influence of rolling direction on the mechanical behavior and formation of cavity stringers in the superplastic Zn-22% Al alloy. *Acta Metall.* **1989**, *37*, 715–723. [CrossRef]
17. Nieh, T.G.; Hsiung, L.M.; Wadsworth, J.; Kaibyshev, R. High strain rate superplasticity in a continuously recrystallized Al–6%Mg–0.3%Sc alloy. *Acta Mater.* **1998**, *46*, 2789–2800. [CrossRef]
18. Pilling, J.; Ridley, N. Effect of hydrostatic pressure on cavitation in superplastic aluminium alloys. *Acta Metall.* **1986**, *34*, 669–679. [CrossRef]
19. Chen, C.L.; Tan, M.J. Cavity growth and filament formation of superplastically deformed Al 7475 Alloy. *Mater. Sci. Eng. A* **2001**, *298*, 235–244. [CrossRef]
20. Conrad, H.; Cao, W.D.; Lu, X.P.; Sprecher, A.F. Effect of electric field on cavitation in superplastic aluminum alloy 7475. *Mater. Sci. Eng. A* **2001**, *138*, 247–258. [CrossRef]
21. Engler, O.; Kuhnke, K.; Hasenclever, J. Development of intermetallic particles during solidification and homogenization of two AA 5xxx series Al-Mg alloys with different Mg contents. *J. Alloy. Compd.* **2017**, *728*, 669–681. [CrossRef]
22. Duan, Y.L.; Xu, G.F.; Xiao, D.; Zhou, L.Q.; Deng, Y.; Yin, Z.M. Excellent superplasticity and deformation mechanism of Al–Mg–Sc–Zr alloy processed via simple free forging. *Mater. Sci. Eng. A* **2015**, *624*, 124–131. [CrossRef]
23. Lee, S.; Utsunomiya, A.; Akamatsu, H.; Neishi, K.; Furukawa, M.; Horita, Z.; Langdon, T.G. Influence of scandium and zirconium on grain stability and superplastic ductilities in ultrafine-grained Al–Mg alloys. *Acta Mater.* **2002**, *50*, 553–564. [CrossRef]
24. Horita, Z.; Furukawa, M.; Nemoto, M.; Barnes, A.J.; Langdon, T.G. Superplastic forming at high strain rates after severe plastic deformation. *Acta Mater.* **2000**, *48*, 3633–3640. [CrossRef]
25. Gashti, S.O.; Fattah-alhosseini, A.; Mazaheri, Y.; Keshavarz, M.K. Effects of grain size and dislocation density on strain hardening behavior of ultrafine grained AA1050 processed by accumulative roll bonding. *J. Alloy. Compd.* **2016**, *685*, 854–861. [CrossRef]
26. Mikhaylovskaya, A.V.; Yakovtseva, O.A.; Golovin, I.S.; Pozdniakov, A.V.; Portnoy, V.K. Superplastic deformation mechanisms in fine-grained Al–Mg based alloys. *Mater. Sci. Eng. A* **2015**, *627*, 31–41. [CrossRef]
27. Cao, F.R.; Li, Z.L.; Zhang, N.X.; Ding, H.; Yu, F.X.; Zuo, L. Superplasticity, flow and fracture mechanism in an Al-12.7Si-0.7Mg alloy. *Mater. Sci. Eng. A* **2013**, *571*, 167–183. [CrossRef]
28. Chokshi, A.H. Cavity nucleation and growth in superplasticity. *Mater. Sci. Eng. A* **2005**, *410–411*, 95–99. [CrossRef]
29. Cao, F.R. *Metal Superplasticity*; Metallurgical Industry Press: Beijing, China, 2014; pp. 211–212, ISBN 978-7-5024-6690-9.
30. Langdon, T.G. A unified approach to grain boundary sliding in creep and superplasticity. *Acta Mater.* **1994**, *42*, 2437–2443. [CrossRef]
31. Mohamed, F.A. Interpretation of superplastic flow in terms of a threshold stress. *J. Mater. Sci.* **1983**, *18*, 582–592. [CrossRef]
32. Raj, R.; Ashby, M.F. Intergranular fracture at elevated temperature Rupture intergranulaire aux temperatures elevees Intergranularer Bruch bei Höheren Temperaturen. *Acta Metall.* **1975**, *23*, 653–666. [CrossRef]
33. Bae, D.H.; Ghosh, A.K. Cavity formation and early growth in a superplastic Al–Mg alloy. *Acta Mater.* **2002**, *50*, 511–523. [CrossRef]
34. Kawasaki, M.; Xu, C.; Langdon, T.G. An investigation of cavity growth in a superplastic aluminum alloy processed by ECAP. *Acta Mater.* **2005**, *53*, 5353–5364. [CrossRef]

35. Chokshi, A.H.; Langdon, T.G. A model for diffusional cavity growth in superplasticity. *Acta Metall.* **1987**, *35*, 1089–1101. [CrossRef]
36. Liu, Z.Y. A new model for cavity nucleation. *Trans. Nonferrous Met. Soc. China.* **1997**, *7*, 145–148.
37. Ghosh, A.K.; Bae, D.H.; Materials, S.L. Initiation and Early Stages of Cavity Growth during Superplastic and Hot Deformation. *Mater. Science Forum.* **1999**, *304–306*, 609–616. [CrossRef]
38. Lee, C.J.; Huang, J.C. Cavitation characteristics in AZ31 Mg alloys during LTSP or HSRSP. *Acta Mater.* **2004**, *52*, 3111–3122. [CrossRef]
39. Takigawa, Y.; Aguirre, J.V.; Taleff, E.M.; Higashi, K. Cavitation during grain-boundary-sliding deformation in an AZ61 magnesium alloy. *Mater. Sci. Eng. A* **2008**, *497*, 139–146. [CrossRef]
40. Mao, W.; Yu, Y. Effect of elastic reaction stress on plastic behaviors of grains in polycrystalline aggregate during tensile deformation. *Mater. Sci. Eng. A* **2004**, *367*, 277–281. [CrossRef]
41. Masuda, H.; Kanazawa, T.; Tobe, H.; Sato, E. Dynamic anisotropic grain growth during superplasticity in Al–Mg–Mn alloy. *Scr. Mater.* **2018**, *149*, 84–87. [CrossRef]

metals MDPI

Article

Microstructural Evaluation and Corrosion Resistance of Semisolid Cast A356 Alloy Processed by Equal Channel Angular Pressing

Mohamed Abdelgawad Gebril [1,2], Mohd Zaidi Omar [1,*], Intan Fadhlina Mohamed [1,3] and Norinsan Kamil Othman [4]

1 Centre for Materials Engineering and Smart Manufacturing, Universiti Kebangsaan Malaysia, 43600 UKM Bangi, Malaysia; gebril.ukm@gmail.com (M.A.G.); intanfadhlina@ukm.edu.my (I.F.M.)
2 Department of Mechanical Engineering, Faculty of Engineering, Benghazi University, 16063 Benghazi, Libya
3 Fuel Cell Institute, Universiti Kebangsaan Malaysia, 43600 UKM Bangi, Malaysia
4 Department of Applied Physics, Faculty of Science and Technology, Universiti Kebangsaan Malaysia, 43600 UKM Bangi, Malaysia; insan@ukm.edu.my
* Correspondence: zaidiomar@ukm.edu.my; Tel.: +603-8921-4831

Received: 21 December 2018; Accepted: 4 February 2019; Published: 8 March 2019

Abstract: As-cast and semisolid casting using a cooling slope A356 alloy were processed by equal channel angular pressing (ECAP) for Si and grain refinement. The ECAP was conducted at room temperature in a mold, with a channel angle of 120°, and this resulted in a significant size reduction of grain and Si particles from 170.5 and 4.22 to 23.12 and 0.71 µm, respectively, after six passes of heat-treated cooling slope casting, using the ECAP process. The hardness increased with ECAP processing, from 61 HV, for the as-cast alloy, to 134 Hv, after six passes of heat-treated cooling slope casting. The corrosion resistance of the alloy improved, from 0.042 to 0.0012 mmy^{-1}, after the ECAP process. In this work both the strength and corrosion resistance of the ECAPed A356 alloys were improved with the application of the cooling slope process than without (i.e., from the as-cast condition).

Keywords: A356 alloy; cooling slope; ECAP; hardness; pitting corrosion

1. Introduction

A356 alloy is an Al–Si casting alloy that contains ~7 wt.% Si, mostly developed for automotive powertrain components such as the engine block and automotive transmission cases. However, the solubility limitation of silicon (Si) in aluminum (Al) contributes to the precipitation of flake-shaped Si particles with sharp edges, which significantly affects the mechanical and electrochemical properties.

The coarse flake Si particles can cause the initiation of premature cracks during deformation. Consequently, this shape weakens the workability of the alloy at room temperature, thus reducing the ductility of the alloy [1]. The morphology of Si particles and its distribution play essential roles in the electrochemical properties of Al–Si alloy. The cathodic behavior of Si within the Al-rich matrix contributes to the occurrence of localized corrosion, with the formation of microgalvanic couples [2,3], which also leads to poor mechanical behavior, such as in stent applications [4]. The reduction in the area ratio of noble Si particles (cathode) to less-noble eutectic Al phase (anode) around Si particles largely improves pitting corrosion resistance. In other words, the reduction in the area ratio of cathode to anode (Ac/Aa) reduces the corrosion current density. In the Al–Si alloy, the size reduction of Si particles which as cathode, facilitates the re-passivation of protective film with improved stability [5–7].

Since the corrosion resistance and strength of alloys are mostly influenced by their microstructure [8–10], there are various methods to refine the microstructure of unmodified A356 alloy in order to improve the corrosion resistance and mechanical properties, such as heat treatment [11–13] and semisolid state processing to obtain a spherical microstructure using mechanical stirring, electromagnetic stirring, and controlled nucleation methods [14–16]. The cooling slope casting process is also considered as another simple method of the semisolid metal casting process to produce feedstock material, with a spheroid microstructure, which bears minimal equipment and operational costs [17–19]. The severe plastic deformation (SPD) technique effectively accommodates the combination of major grain refinement and Hall-Petch strengthening in bulk billets, where the working piece is subjected to extensive strains of cold or warm processing. Fundamentally, the SPD technique is eminently capable of effectively refining the structure of grain down to an ultrafine micrometer-scale or nano-scale due to the agglomeration of dislocation, which leads to the formation of subgrains. In particular, this technique enables various metals and metallic alloys that are brittle and ductile to achieve the structure of refined grain [20–22].

The equal channel angular pressing (ECAP) of SPD the technique is considered the most effective technique in fabricating bulk ultrafine-grained materials [23]. Some researchers have applied several methods of SPD to Al–Si alloy to improve the corrosion resistance of the alloy through microstructural refinement. In one study, [24,25] it was concluded that the improvement of the corrosion resistance and mechanical properties of pure aluminum and A356 alloy was due to microstructure refinement. In another study, the effect of ECAP processing on the corrosion resistance of Al—11wt.% Si [26], it was reported that the great refinement of both the matrix microstructure and the Si particles after severe plastic deformation leads to an improvement of the corrosion resistance.

In another research [27] on the Anticorrosion behavior of ultrafine-grained Al—26 wt.% Si alloy fabricated by ECAP, it was reported that the improvement of ECAPed Al—26 wt.% Si alloy corrosion resistance results from the homogeneous ultrafine structure, with the breakage of brittle large primary silicon crystals. SPD processing improves the corrosion resistance and mechanical properties of Al–Si alloy [28].

However, the reduction of both grain size and eutectic Si particles through the deformation process may alter the corrosion property of the alloy [10,29,30]. Addressing these issues, the present study aimed to refine Si particles and to produce material with a uniform microstructure through the combination of the cooling slope casting process and ECAP processing at room temperature. The effect of T6 heat treatment was studied to determine the microstructure changes after ECAP processing. This study further examined the effects of microstructure changes and its effect on strength as well as corrosion resistance for both ECAPed as-cast and ECAPed cooling slope-cast A356 alloy.

2. Materials and Methods

This study used cast commercial A356 (Si—7 wt.%, Mg—0.149 wt.%, Fe—0.126 wt.%, Cu—0.01 wt.%, Mn—0.002 wt.%, Zn—0.006 wt.%, Cr—0.001 wt.%, Ti—0.178 wt.%) alloy casted in ingot, with initial dimensions of 80 mm × 40 mm × 140 mm (width × thickness × length). Using a graphite crucible, the as-cast was melted at a temperature of 750 °C. For the cooling slope casting process, the cooling slope of stainless steel, with a slope length of 250 mm as well as a tilt angle of 60°, was used. The reason for selecting these conditions was based on our work [31]. The apparatus used for this process is shown in Figure 1a. This study specifically selected 620 °C as the pouring temperature to limit the superheating of the melt [31]. The molten metal was poured downward, on a stainless steel slope, into a mold with a vertical surface, before quenching in water. In line with the T6 procedure, this study performed a heat treatment process to the as-cast and cooling slope-cast samples, which involved the following processes in this order: (1) an eight-hour sequence of solution treatment at 540 °C, (2) water quenching, and (3) three-hour aging process at a temperature of 180 °C [32]. Following that, as illustrated in Figure 1b, as-cast and cooling slope-cast samples were machined into a rod shape, with a diameter of 9.8 mm. The samples were ECAPed in a die with a channel angle of 120°, following route A (where the sample is not rotated between each pass). It should be noted that molybdenum disulfide (MoS_2) grease was used as a lubricant in this study.

The pressing of as-cast and cooling slope-cast A356 alloy samples was carried out using a 50-ton hydraulic press. As-cast samples were pressed up to four passes, without any cracks on the surface while the cooling slope casting samples were successfully pressed to six passes.

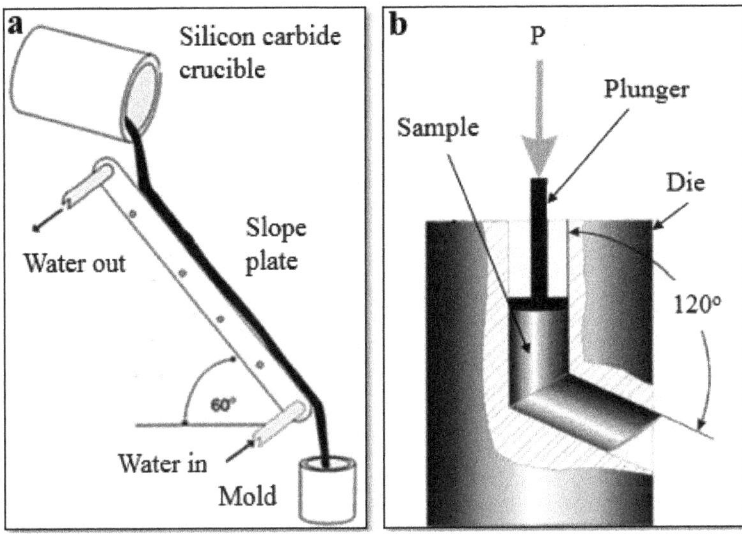

Figure 1. Schematic of (**a**) cooling slope casting and (**b**) ECAP mold.

Both ECAPed samples were subsequently examined under a field emission scanning electron microscope (FESEM, Zeiss, Oberkochen, Germany) and optical microscope (OM, Olympus corporation, Tokyo, Japan). Additionally, an energy dispersive x-ray (EDX) (equipped to FESEM) was used for elemental analysis. A Vickers hardness tester (micro Vickers hardness tester, Zwick, Germany; ZHVμ) was used to measure the hardness of the average of three samples per case. These samples were also prepared for microstructure analysis using silicon carbide (SiC) papers of grit between 180 and 2000, followed by a polishing process using 3 μm and 1 μm of diamond paste (Al_2O_3). Meanwhile, an etching process was performed using Keller's reagent (1% HF, 1.5% HCl, 2.5% HNO_3, H_2O solution) as an etchant. This study performed quantitative metallography analysis to measure the grain size, according to the ASTM E112 standard. The size of Si particles (the width and length of particles) was measured using the Smart Tiffv2 software, considering at least 200 particles in each case. Following that, an electrochemical experiment was performed in naturally aerated 3.5% NaCl solution at room temperature, with pH 6.5. A potentiostat GAMRY 3.2 was used to measure the rate of corrosion (characterized by i_{corr}) of these samples. For this, a three-electrode cell was used, which comprised of (1) test material (as a working electrode), (2) graphite (as a counter electrode), and (3) a silver or silver chloride (Ag or AgCl) electrode (as a reference electrode). The potential dynamic polarization tests were performed at a scanning rate of 1 mV s^{-1} with a range from −250 mV versus open circuit potential (OCP) to the final potential of 250 mV versus OCP. The potentiodynamic tests were started after about 15 min of immersion in 3.5% NaCl. Immersion tests were performed during 14 days in 3.5 wt.% NaCl naturally aerated solution to study the surface appearance. Basically, each sample was mounted in epoxy that was aired for 24 hours. Finally, these samples were smoothened using up to 1200-grit SiC before each corrosion test.

3. Results and Discussion

3.1. The Effect of Cooling Slope

Figure 2 shows the microstructure of the as-cast A356 alloy sample, before and after the cooling slope casting process. Typically, the primary α-Al phase (bright phase), which is surrounded by eutectic phase (dark phase), is formed during the initial solidification phase.

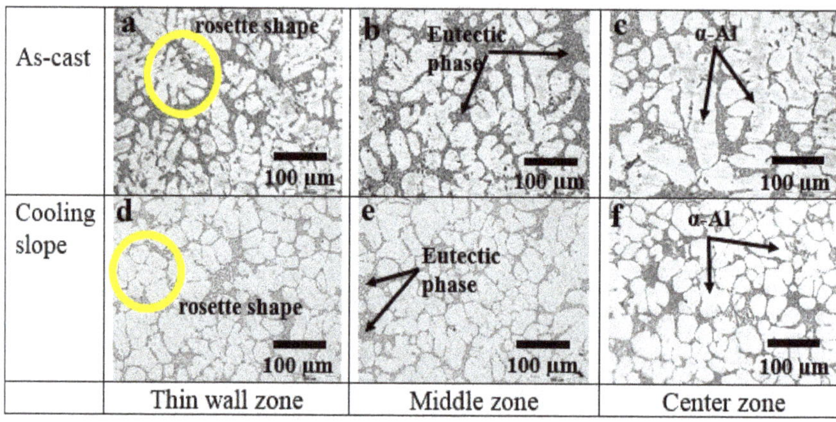

Figure 2. Microstructure of as-cast (**a–c**) and cooling slope casting samples (**d–f**). (**a**) coarse dendritic and rosette shape of α-Al phase at thin wall zone; (**b**) and (**c**) dendritic shape in the middle and center zone in as-cast samples; (**d**) rosette shape of α-Al phase at thin wall zone; (**e**) and (**f**) nearly globular shape at the middle and in the center zone of cooling slope samples.

The morphology of primary α-Al phase in three zones was found to be mostly dendritic for the as-cast sample, before the cooling slope casting process, as displayed in Figure 2a–c. However, the evolution of the coarsen dendritic to finer dendritic could be observed from the center zones approaching the thin wall zone. The variation in cooling rate depends on the location in the mold and affects the morphology of the primary α-Al phase. Considering that the mold itself has a relatively low initial temperature, the cooling rate is more rapid near the wall of the mold, thus leading to the nucleation of numerous grains of random orientations in the thin wall zone area [33]. During the pouring of molten alloy, the presence of forced convection detaches the dendritic arms that developed in the thin wall zone area. However, the dendrites start to form in the middle zone due to the lower cooling rates in this zone. They later grew and become ripened as coarse dendrites as a result of the lowest cooling rate in this center zone [34].

The non-dendritic microstructures of the as-cast A356 sample, after the cooling slope casting process, are shown in Figure 2d–f. However, in particular, the morphology of the primary α-Al phase in the thin wall zone appeared rosette-like Figure 2d and almost globular-like in the middle zone and center zone Figure 2e–f. In the center zone, the primary α-Al phase was found to be coarser than that in the middle zone. This may be because the temperature gradient across these three sections during solidification contributes to the variation in the morphology and size of the primary α-Al phase in these three zones. The slowest cooling rate in the center zone allows the α-Al phase to grow courser, as the cooling time is prolonged.

3.2. The Heat Treatment Solution

Alloy A356 is a heat-treatable hypoeutectic Al–Si alloy. Figure 3a–b shows optical micrographs of the both the primary α-Al phase and the eutectic Si particles in both alloy samples, before the heat treatment process. As for the primary α-Al phase (bright phase), grains with a globular structure could be observed in the cooling slope-cast alloy sample, while large grain, surrounded by coarse eutectic constituents (dark phase), appeared in the as-cast alloy sample. The flake and acicular morphology of Si particles in as-cast and in cooling slope samples were transformed into a spheroidized shape after T6 heat treatment, as shown in Figure 3c–d. In a solution heat treatment sample, the eutectic Si particles were fragmented and spheroidized during the coarsening processes [35,36]. It was observed that the eutectic Si particles were refined and had fewer sharp angles edges, as in the as-cast alloy before heat treatment. The same behavior was reported by previous studies, which found that the T6 heat treatment process initiated the spheroidization of Si particles [37,38], but a considerable amount of acicular shapes remained after the T6 heat treatment process.

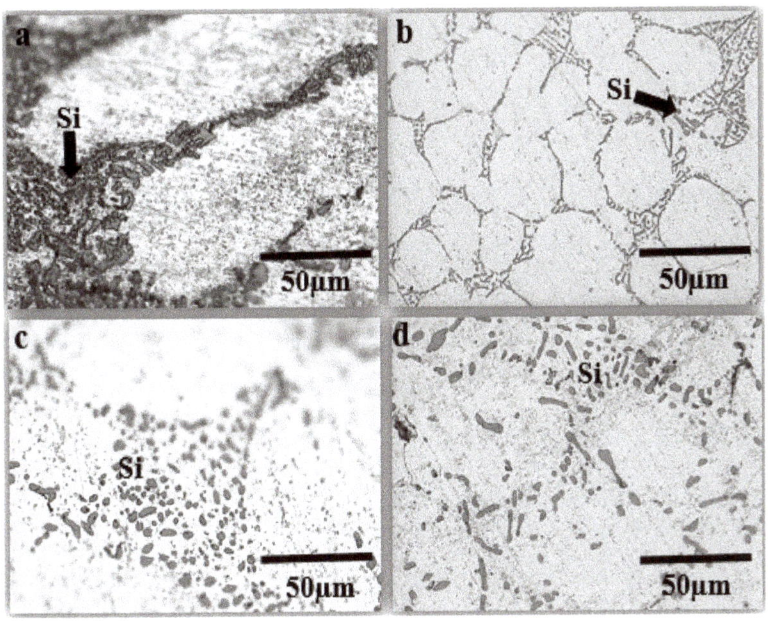

Figure 3. Microstructure of eutectic phase: (**a**) as-cast, (**b**) cooling slope, (**c**) as-cast-T6 and (**d**) cooling slope-T6.

3.3. Process of ECAP

The microstructures of the as-cast-T6 and cooling slope-cast-T6 alloy samples, subjected to four passes of route A, were displayed in Figure 4a–b. Meanwhile, following route A, the microstructure of cooling slope-cast-T6 alloy samples, which was subjected to six passes of ECAP, is shown in Figure 4c. The primary α-Al phase and eutectic constituents were elongated into plate-like shapes for the as-cast alloy sample [39], but fibrous-like shapes, for the latter samples. As shown in Table 1, the Si particles and eutectic phase appeared to be finer in both as-cast and cooling slope-cast alloy samples, after ECAP processing. Due to ECAP processing, the distribution was observed to be more uniform in the cooling slope-cast alloy sample. The result determined agrees with our earlier study, which reported that refine Si particles were observed through the cooling slope casting of alloy [40]. Nonetheless, the spheroidization and fragmentation of Si particles in this study were acquired through the T6 heat treatment process, coupled with the ECAP process.

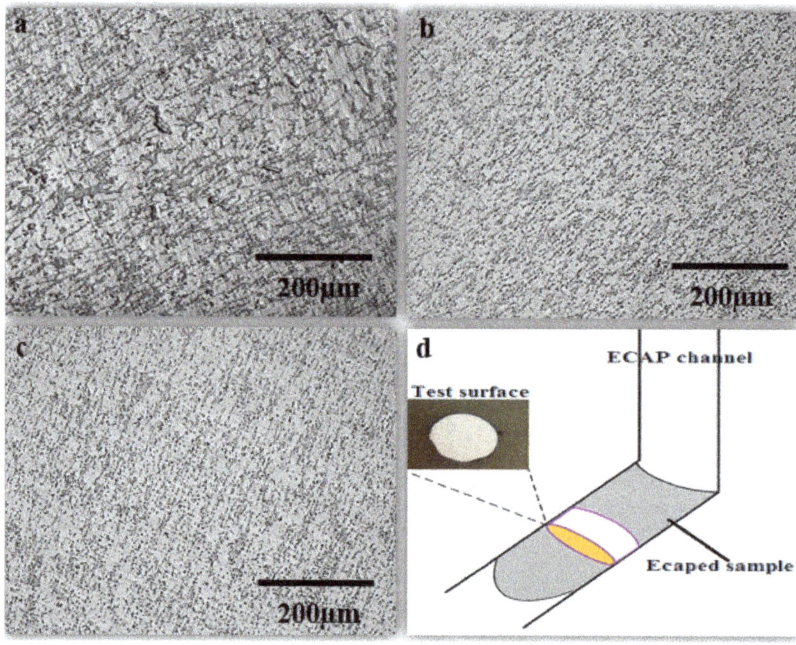

Figure 4. Microstructure of ECAPed-T6 A356 alloy (**a**) as-cast after 4 passes, (**b**) cooling slope after 4 passes, (**c**) cooling slope after 6 passes and (**d**) surface of microscopic ECAPed sample.

Based on the microstructures of both samples, following route A, the refining and the distribution of both primary the α-Al phase and eutectic constituents in the ECAPed cooling slope-cast sample became more homogeneous. Thus, the microstructure changes may lead to changes in the mechanical and electrochemical properties of the alloy [28,41–43]. This is due to the reduction and distribution of cathodic to anodic phases. The straining led to the formation of dislocated cell structures, with a high dislocation density. The dislocation remains within the cell structures, which do not affect the remaining cell boundaries or develop walls that separate these cells into smaller cells. However, persistent shear straining processes may lead to a saturated dislocation density within these cell structures, which could be significantly reduced through (i) enhanced dynamic recovery (which would stabilize the creation and annihilation of dislocations) and (ii) the conversion of cells to well-defined grains (which would cause severe movement of cell interior dislocations to cell boundaries). In fact, this suggested that the process of grain refinement depends on the level of straining or in other words, the generation of dislocations [44–46].

3.4. Scanning Electron Microscope (FESEM)

Table 1 shows the grain refinement and Si particles fragmentation after ECAP processing. Figure 5a–b shows the scanning electron microscope (SEM) micrographs of the ECAPed as-cast and cooling slope-cast samples. The Si particles appeared finer in the ECAPed cooling slope than in the ECAPed as-cast. With the continuous passes, the Si particles and eutectic mixture were broken down and fragmented into finer particles, as shown in Figure 5c–d.

Table 1. Average size of grain size and Si particles before and after ECAP.

Samples	Si Size (μm)	Grain Size (μm)
As-cast	4.22	170.51
ECAPed as-cast-T6, 4 passes	0.761	40.40
Cooling slope	3.01	53.55
ECAPed cooling slope-T6, 4 passes	0.74	29.34
ECAPed cooling slope-T6, 6 passes	0.71	23.12

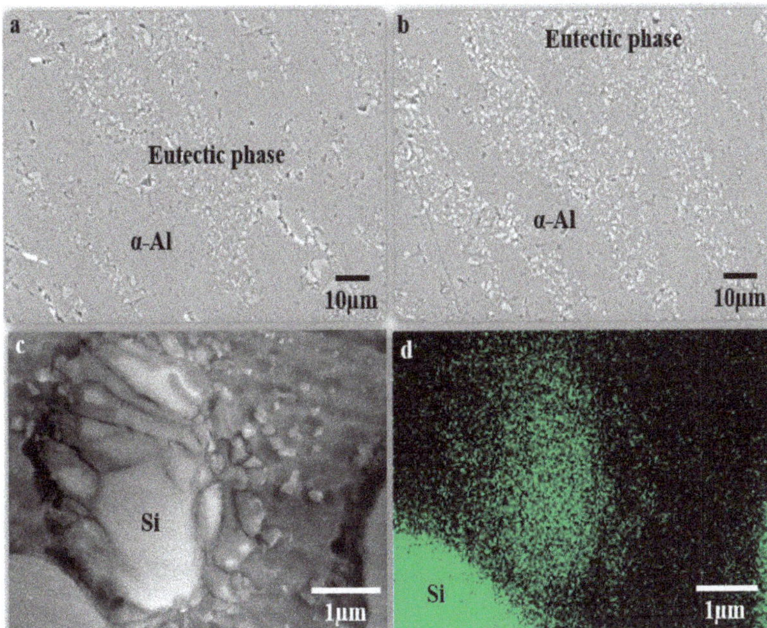

Figure 5. FESEM images: Si particles morphology after ECAP (**a**) as-cast, (**b**) cooling slope, (**c**) and (**d**) SEM and mapping of Si fragmentation of ECAPed sample.

3.5. Hardness

Figure 6a shows the Vickers microhardness of as-cast and cooling slope samples, measured after a combination of heat treatment and ECAP processing via route A. It shows that the hardness increased from 61 HV to 125 and 129 Hv, after four passes of ECAPed as-cast and cooling slope samples, respectively. After six passes, the as-cast sample failed to reach six passes, without surface cracks, while the cooling slope sample successfully reached six passes, with a 134 HV microhardness. After heat treatment, the microhardness was enhanced for the as-cast and cooling slope samples by 26% and 36%, respectively.

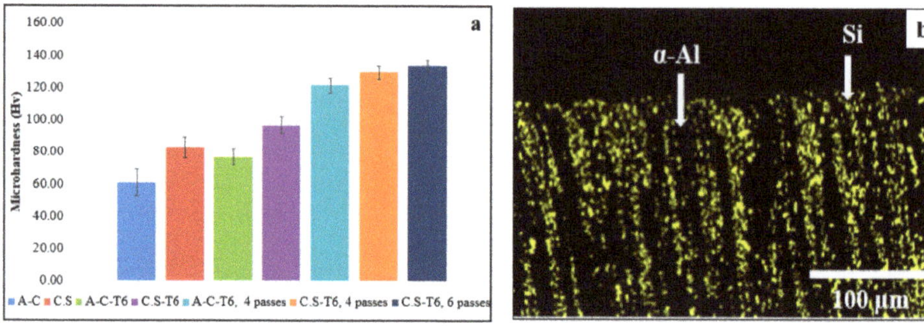

Figure 6. (**a**) Effect of T6 heat treatment and ECAP process on hardness of A356 alloy and (**b**) Mapping of silicon particles distribution in ECAPed samples.

The spheroidization of eutectic Si after T6 heat treatment was found to increase the hardness of samples. Essentially, the spheroidization of Si particles after T6 heat treatment, and the precipitation of magnesium silicide (Mg_2Si) particles during the aging process, tend to increase the ultimate tensile strength as well as hardness [47,48].

Since shear force could break down the dendrite arms of the α-Al phase, leading to grain refinement, its microstructure in the rheocasting condition became smaller and denser, where the rheocast sample recorded the highest microhardness, compared to the as-cast sample [12], as well as a change in the morphology of Si particles, from a flake shape in the as-cast to acicular shapes in the cooling slope, which contributed to the microhardness of A356 alloy, as mentioned in Section 3.2.

Four passes of ECAPed-T6 for the as-cast and cooling slope casting samples increased the hardness of both samples due to the fragmentation of globular heat-treated eutectic Si particles, the reduction in grain size, and increase in the density of dislocations. High strain, induced through six passes within the cooling slope sample during the process of ECAP, increases both the dislocation density and grain refinement as well as the fragmentation of eutectic Si particles, which led to a greater improvement in hardness, which is in line with other studies [39,49,50].

The homogenous distribution of fragmented eutectic Si particles and primary α-Al phase of the cooling slope-cast sample plays an integral role in improving the hardness of ECAPed materials. As shown in Figure 4c, the homogeneity of the distribution of the primary α-Al phase and Si particles within the eutectic mixture phase, for the microstructure of ECAPed cooling slope-cast sample, surpassed the homogeneity of a similar distribution of the ECAPed as-cast sample. Figure 6b displays the EDX map of the distribution of Si particles, after the ECAP process for the cooling slope sample. Due to the microstructural evolution, after the processes of semisolid casting, coupled with ECAP, it was expected that the finer formation and homogeneous distribution of α-Al and eutectic mixture phase enhanced the mechanical properties of the material [51].

3.6. Corrosion Resistance

3.6.1. Surface Morphology

Figure 7 reveals optical micrographs of the surface morphology, before and after the processes of ECAP, after immersion in 3.5% NaCl solution for ten days, for both the as-cast and cooling slope-cast samples. Overall, the size and the number of pitting corrosions, as well as large corrosion rings products around pits, before and after the process of ECAP, for the cooling slope-cast sample, were found to be lesser than those for the as-cast sample due to reduction and redistribution of cathodic phases after the ECAP process. Larger localized corrosion pits were found to be noticeable, for the as-cast sample. The surface morphology indicated the formation of stable pitting [52], which was attributed to a localized corrosion attack between the active particles and noble particles in a eutectic phase [53]. Essentially, the non-presence of corrosion products within the corrosion rings indicates that the cathodic reaction occurs on rings, while the anodic reaction occurs inside the stable pit instead. Applied strain during the ECAP process reduces the grain size and develops crystalline defects, including dislocations of the grain boundary. An increase in both the area of grain boundaries and dislocations led to the formation of passive films and the corrosion of ultrafine grains, improved by the rapid formation of passive films at surface crystalline defects, including grain boundaries and dislocations [54,55]. In Al alloy, Al oxide film, containing eutectic Si particles, improved the pitting corrosion resistance by increasing the ECAP pass number, which was related to the decrease of the size of Si-containing impurities, because Si is the major cause of pitting corrosion. The decreases of the cathodic area led to a consequent decrease of the anodic current density [56].

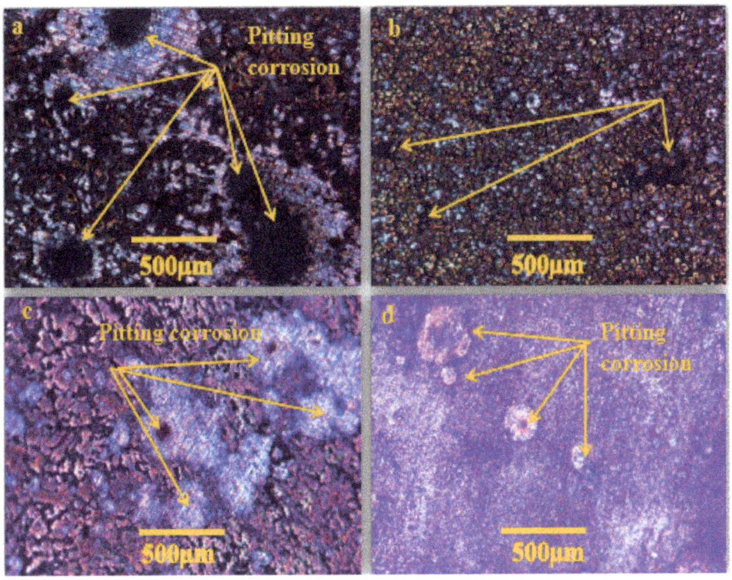

Figure 7. Surface appearance of A356 alloy (**a**) as-cast, (**b**) cooling slope (**c**) ECAPed as-cast-T6 after 4 passes and (**d**) ECAPed cooling slope-T6 after 6 passes after immersion for 10 days.

3.6.2. Potentiodynamic Test

The electrochemical behavior of A356 alloy was evaluated through the exposure of these samples to a corrosive environment, based on a simulation of sea water, using 3.5% NaCl electrolyte solution at room temperature [57]. The rate of corrosion was measured using the linear polarization technique through the Tafel extrapolation method to identify the corrosion resistance of both as-cast and cooling slope-cast samples. To compare these samples, Figure 8a–c depicts the polarization curves for the as-cast and cooling slope-cast samples, before and after the ECAP process. The corrosion performance of A356 alloy in 3.5 wt.% NaCl solution are tabulated in Table 2.

Based on the curves presented, the estimated average corrosion potentials were found to be approximately similar, with trivial differences. The results further revealed that the reduction in the corrosion rate and increment of polarization resistance, after the T6 heat treatment process for both as-cast and cooling slope-cast samples, could be attributed to the modification in the shape of certain Si particles, where these particles became substantially finer after the processes of ECAP and cooling slope casting [40]. Therefore, this could be associated with a reduction in the area ratio of cathodic to anodic phases. The corrosion rate of the as-cast alloy was 0.042 mmpy, reduced to 0.0015 mmpy after T6-4 passes route of A. Additionally, the corrosion rate of the cooling slope casting alloy was 0.019 mmpy, reduced to 0.0014 and 0.00125 mmpy after 4 and 6 passes, respectively.

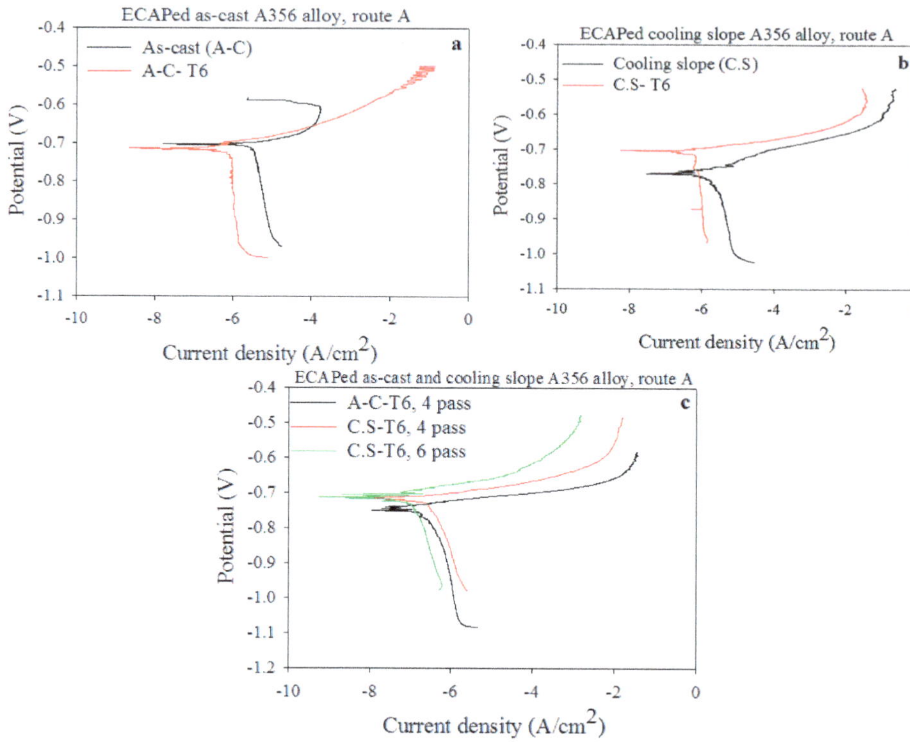

Figure 8. Polarization curves of A356 alloy before and after ECAP in 3.5 wt.% NaCl. (**a**) as-cast before and after T6; (**b**) cooling slope before and after T6; (**c**) heat-treated T6, as-cast and cooling slope after 4 and 6 passes.

Essentially, the polarization resistance depends on the microstructural state. After the ECAP process, the polarization resistance for both samples increased with more ECAP passes. Nevertheless, after the ECAP process, the polarization resistance for the as-cast sample was found to be inferior to that of cooling slope-cast sample, which positively affected the reconstruction of the metal protective layer, as shown in Table 2. However, the rate of corrosion for the heat-treated as-cast and cooling slope-cast samples decreased with more ECAP passes.

The cooling slope-cast sample mainly demonstrated an exceptional corrosion resistance, which was greater than that of the as-cast sample. The fine-grained structure, with higher grain boundaries, reduces the concentration of chloride per grain boundary, which reduces the current density [58,59], and this provides the advantage of forming more a stable and intact passivation film, improving the corrosion resistance. The obtained result in this study is in line with [60–62].

Table 2. Average of corrosion rate (CR), polarization resistance (Rp) and current density (I_{corr}).

Samples	E_{corr} (V)	I_{corr} (A/cm^2)	Rp ($\Omega\cdot m^2$)	β_c (V·dec^{-1})	β_a (V·dec^{-1})	CR (mmy^{-1})
As-cast (A-C)	−0.698	3.894×10^{-6}	5.212×10^3	0.642	0.0504	0.0424
A-C-T6	−0.713	8.432×10^{-7}	2.035×10^4	0.443	0.0499	0.0160
A-C-T6, 4 passes	−0.751	1.369×10^{-7}	9.147×10^4	0.268	0.0286	0.0015
Cooling slope (C.S)	−0.769	1.790×10^{-6}	9.690×10^3	0.333	0.0480	0.0195
C.S-T6	−0.702	7.258×10^{-7}	2.180×10^4	0.321	0.0474	0.0079
C.S-T6, 4 passes	−0.717	1.285×10^{-7}	8.516×10^4	0.280	0.047	0.0014
C.S-T6, 6 passes	−0.709	1.145×10^{-7}	9.525×10^4	0.259	0.0466	0.00124

3.6.3. Pitting Corrosion Appearance

In the eutectic mixture phase, the initial corrosion stages of the as-cast sample revealed localized corrosion, whereas the silicon particles and α-Al interfaces were unaffected [8,32,33,49]. Figure 9 reveals a side view of the pitting corrosion of A356 alloy. Figure 9a,b displays the pitting corrosion of (a) the as-cast and (b) cooling slope-cast of A356 alloy samples. It should be noted that the presence of the enlarged area in the eutectic mixture phase in Figure 9a,b would eventually lead to a wider area of corrosion on the surface of alloy.

This implied that the microgalvanic corrosion between impurities of high Si and Al matrix could contribute to the occurrence of pitting corrosion within the area that contains impurities of high Si. Meanwhile, Figure 9c,d displays the pitting corrosion, after a potentiodynamic test for the heat-treated (c) as-cast sample (after four passes of ECAP process) and (d) cooling slope-cast sample (after six passes of ECAP process). The difference in the area of the α-Al separated grains and depth of corrosion in the eutectic mixture phase, between both the as-cast and cooling slope-cast samples, before and after the ECAP process, was evident, as revealed in Figure 9a,b. The reduction in the size of Si particles, as previously shown in Figure 5 and Table 1, the refined grain in an elongated shape, as well as the microstructure homogeneity of eutectic mixture phase surrounding the elongated α-Al phase led to the reduced rate of corrosion for the ECAPed A356-T6 cooling slope-cast alloy sample, compared to that of the as-cast alloy sample, as displayed in Figure 9c,d.

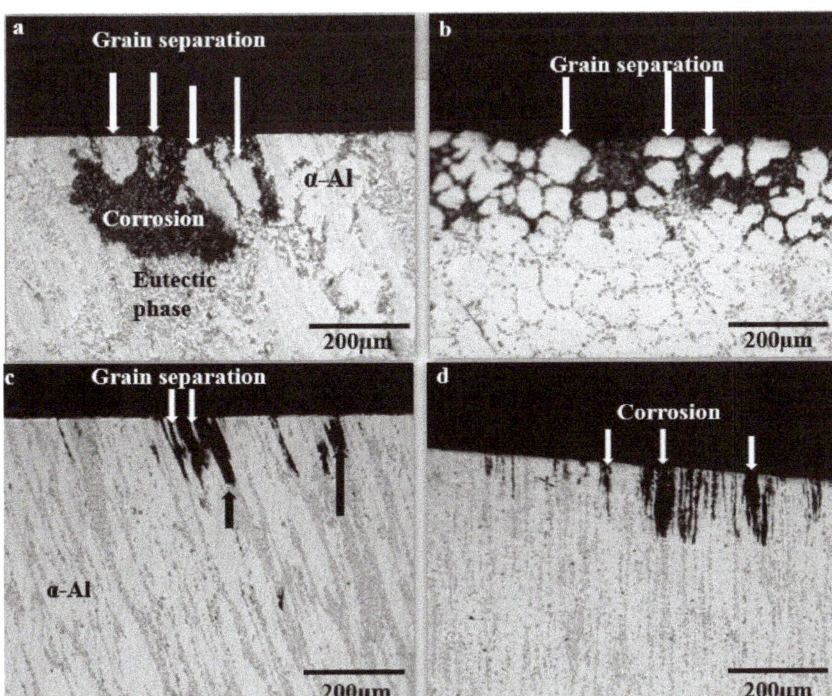

Figure 9. Pitting corrosion of (**a**) as-cast, (**b**) cooling slope, (**c**) as-cast-T6,4 passes and (**d**) cooling slope-T6, 6 passes route A of A356 alloy.

In summary, the corrosion resistance of ECAPed A356-T6 alloy sample was significantly improved due to the reduction in the galvanic potential difference, which the reduced area ratio of the noble phase to less-noble eutectic mixture phase contributed to.

The shape of Si particles influences the formation of a firm oxide film. In particular, coarsened Si particles impede the development of a protective oxide layer in the aqueous solutions and weakens the protective passive film [8]. The reduction in the area ratio of noble Si particles to less-noble eutectic Al phase around these Si particles significantly improves the resistance of pitting corrosion [5]. Furthermore, localized mechanical damage or the chemical changes of the environment may damage the protective oxide film. The increase of the applied voltage on the side of the cathodic curve, as shown in Figure 8, contributes to the rapid reduction of current to the extent that its value remains unchanged before achieving the value of I_{corr}.

The steep increment in the anodic current, with the increment in the applied potential in the positive direction, breaks the developed oxide layer and causes pitting corrosion. Here, the chloride ion attacks and dissolves the aluminum surface to form an aluminum chloride compound. The SEM image and EDX of the Al_2O_3 oxide layer of the ECAPed A356 alloy after the corrosion test, are shown in Figure 10. The highest corrosion resistance with the lowest corrosion rate were obtained for the aluminum A356 alloy, which was subjected to the ECAP process, as tabulated in Table 2.

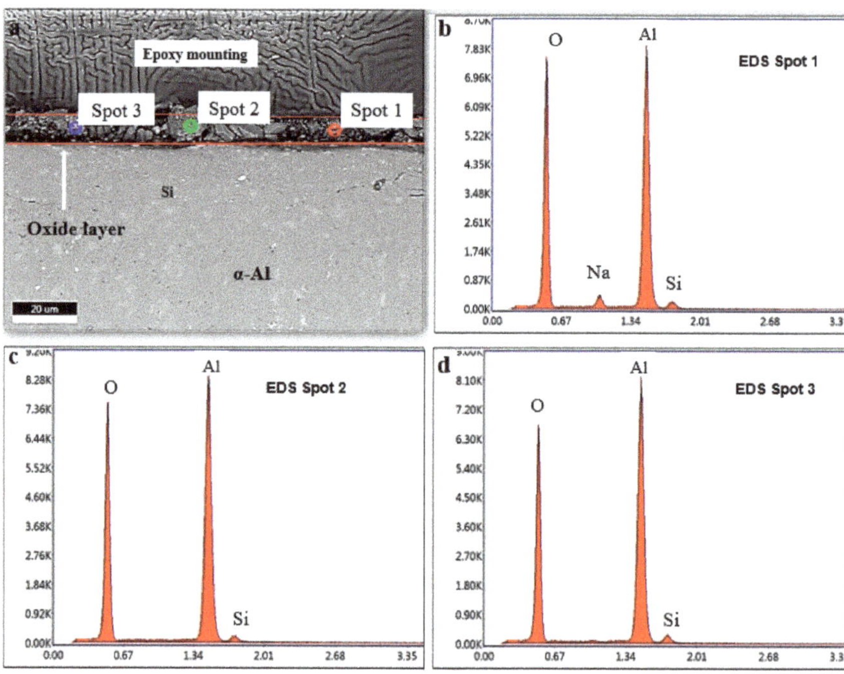

Figure 10. (**a**) Oxide layer of ECAPed A356 alloy and (**b–d**) EDX of oxide layer.

4. Conclusions

In this study, the microstructural evaluation of A356 alloy using heat treatment and cooling slope followed by ECAP process was successfully carried out. The effect of grain refinement on the hardness and the corrosion resistance was investigated and the results obtained can be summarized as follows:

- The as-cast samples were successfully subjected to ECAP for up to four passes, while the samples from cooling slope casting, six passes.
- The latter also showed finer and more homogeneous distribution of α-Al grains and Si particles. As a result, their hardness values were also higher.
- The combination of cooling slope casting and ECAP had given the lowest current density to the A356 alloy, at 1.145×10^{-7} A/cm^2 when experimental using 3.5 wt.% NaCl solution. The favorable

corrosion resistance was attributed to the refined Si particles that impeded the occurrences of microgalvanic cells on the protective layer of the alloy surface.

Author Contributions: M.A.G. carried out the experimental works, analysis and writing under the supervision of M.Z.O., I.F.M. and N.K.O.

Funding: This research was funded by Ministry of Education Malaysia and Universiti Kebangsaan Malaysia [DIP-2016-007].

Acknowledgments: The authors wish to express their gratitude to the Ministry of Education Malaysia and Universiti Kebangsaan Malaysia for the financial assistance and support.

Conflicts of Interest: The authors declare no conflict of interest.

References

1. Haghdadi, N.; Zarei-Hanzaki, A.; Abedi, H.R.; Sabokpa, O. The effect of thermomechanical parameters on the eutectic silicon characteristics in a non-modified cast A356 aluminum alloy. *Mater. Sci. Eng. A* **2012**, *549*, 93–99. [CrossRef]
2. Davis, J.R. Aluminum and Aluminum Alloys. *Light Met. Alloy* **2001**, *66*, 351–416. [CrossRef]
3. Musa, A.Y.; Mohamad, A.B.; Kadhum, A.A.H.; Chee, E.P. Galvanic corrosion of aluminum alloy (Al2024) and copper in 1.0 M nitric acid. *Int. J. Electrochem. Sci.* **2011**, *6*, 5052–5065.
4. Galvin, E.; O'Brien, D.; Cummins, C.; Mac Donald, B.J.; Lally, C. A strain-mediated corrosion model for bioabsorbable metallic stents. *Acta Biomater.* **2017**, *55*, 505–517. [CrossRef] [PubMed]
5. Tahamtan, S.; Fadavi Boostani, A. Evaluation of pitting corrosion of thixoformed A356 alloy using a simulation model. *Trans. Nonferrous Met. Soc. China English Ed.* **2010**, *20*, 1702–1706. [CrossRef]
6. Boostani, A.F.; Tahamtan, S. Fracture behavior of thixoformed A356 alloy produced by SIMA process. *J. Alloys Compd.* **2009**, *481*, 220–227. [CrossRef]
7. Tahamtan, S.; Boostani, A.F. Quantitative analysis of pitting corrosion behavior of thixoformed A356 alloy in chloride medium using electrochemical techniques. *Mater. Des.* **2009**, *30*, 2483–2489. [CrossRef]
8. Barbucci, A.; Bruzzone, G.; Delucchi, M.; Panizza, M.; Cerisola, G. Breakdown of passivity of aluminium alloys by intermetallic phases in neutral chloride solution. *Intermetallics* **2000**, *8*, 305–312. [CrossRef]
9. Wang, Q.G. Microstructural effects on the tensile and fracture behavior of aluminum casting alloys A356/357. *Metall. Mater. Trans. A* **2003**, *34*, 2887–2899. [CrossRef]
10. Arrabal, R.; Mingo, B.; Pardo, A.; Mohedano, M.; Matykina, E.; Rodríguez, I. Pitting corrosion of rheocast A356 aluminium alloy in 3.5 wt.% NaCl solution. *Corros. Sci.* **2013**, *73*, 342–355. [CrossRef]
11. Akhter, R.; Ivanchev, L.; Burger, H.P. Effect of pre/post T6 heat treatment on the mechanical properties of laser welded SSM cast A356 aluminium alloy. *Mater. Sci. Eng. A* **2007**, *447*, 192–196. [CrossRef]
12. Liao, B.; Park, Y.; Ding, H. Effects of rheocasting and heat treatment on microstructure and mechanical properties of A356 alloy. *Mater. Sci. Eng. A* **2011**, *528*, 986–995. [CrossRef]
13. Syarif, J.; Detak, Y.P.; Ramli, R. Modeling of Correlation between Heat Treatment and Mechanical Properties of Ti–6Al–4V Alloy Using Feed Forward Back Propagation Neural Network. *ISIJ Int.* **2010**, *50*, 1689–1694. [CrossRef]
14. Atkinson, H. Current status of semi-solid processing of metallic materials. *Adv. Mater. Form.* **2007**, 81–98. [CrossRef]
15. Mohammed, M.N.; Omar, M.Z.; Salleh, M.S.; Alhawari, K.S.; Kapranos, P. Semisolid metal processing techniques for nondendritic feedstock production. *Sci. World J.* **2013**, *2013*, 1–16. [CrossRef] [PubMed]
16. Salleh, M.S.; Omar, M.Z.; Syarif, J.; Mohammed, M.N. An Overview of Semisolid Processing of Aluminium Alloys. *ISRN Mater. Sci.* **2013**, *2013*, 1–9. [CrossRef]
17. Spencer, D.B.; Merhrabian, R.; Flemings, M. Rheological behaviour of Sn-15pct Pb in the crystallization range. *Metall. Trans.* **1972**, 3, 1925–1932. [CrossRef]
18. Haga, T.; Kapranos, P. Billetless simple thixoforming process. *J. Mater. Process. Technol.* **2002**, *130–131*, 581–586. [CrossRef]
19. Birol, Y. A357 thixoforming feedstock produced by cooling slope casting. *J. Mater. Process. Technol.* **2007**, *186*, 94–101. [CrossRef]

20. Mohamed, I.F.; Masuda, T.; Lee, S.; Edalati, K.; Horita, Z.; Hirosawa, S.; Matsuda, K.; Terada, D.; Omar, M.Z. Strengthening of A2024 alloy by high-pressure torsion and subsequent aging. *Mater. Sci. Eng. A* **2017**, *704*, 112–118. [CrossRef]
21. Fritsch, S.; Wagner, M.F.-X. On the effect of natural aging prior to low temperature ECAP of a high-strength aluminum alloy. *Metals* **2018**, *8*, 63. [CrossRef]
22. Wei, J.; Huang, G.; Yin, D.; Li, K.; Wang, Q.; Zhou, H. Effects of ECAP and Annealing Treatment on the Microstructure and Mechanical Properties of Mg-1Y (wt.%) Binary Alloy. *Metals* **2017**, *7*, 119. [CrossRef]
23. Shin, D.H.; Kim, B.C.; Park, K.T.; Choo, W.Y. Microstructural changes in equal channel angular pressed low carbon steel by static annealing. *Acta Mater.* **2000**, *48*, 3245–3252. [CrossRef]
24. Abd El Aal, M.I.; Sadawy, M.M. Influence of ECAP as grain refinement technique on microstructure evolution, mechanical properties and corrosion behavior of pure aluminum. *Trans. Nonferr. Met. Soc. China Engl. Ed.* **2015**, *25*, 3865–3876. [CrossRef]
25. Moradi, M.; Nili-Ahmadabadi, M.; Heidarian, B. Improvement of mechanical properties of AL (A356) cast alloy processed by ecap with different heat treatments. *Int. J. Mater. Form.* **2009**, *2*, 85–88. [CrossRef]
26. Jiang, J.-H.; Ma, A.-B.; Lu, F.-M.; Saito, N.; Watazu, A.; Song, D.; Zhang, P.; Nishida, Y. Improving corrosion resistance of Al-11mass% Si alloy through a large number of ECAP passes. *Mater. Corros.* **2011**, *62*, 848–852. [CrossRef]
27. Jiang, J.; Ma, A.; Song, D.; Yang, D.; Shi, J.; Wang, K.; Zhang, L.; Chen, J. Anticorrosion behavior of ultrafine-grained Al-26 wt% Si alloy fabricated by ECAP. *J. Mater. Sci.* **2012**, *47*, 7744–7750. [CrossRef]
28. Wang, X.; Nie, M.; Wang, C.T.; Wang, S.C.; Gao, N. Microhardness and corrosion properties of hypoeutectic Al-7Si alloy processed by high-pressure torsion. *Mater. Des.* **2015**, *83*, 193–202. [CrossRef]
29. Osório, W.R.; Garcia, L.R.; Goulart, P.R.; Garcia, A. Effects of eutectic modification and T4 heat treatment on mechanical properties and corrosion resistance of an Al-9 wt% Si casting alloy. *Mater. Chem. Phys.* **2007**, *106*, 343–349. [CrossRef]
30. Tahamtan, S.; Fadavi Boostani, A. Microstructural characteristics of thixoforged A356 alloy in mushy state. *Trans. Nonferr. Met. Soc. China Engl. Ed.* **2010**, *20*, s781–s787. [CrossRef]
31. Samsudin, M.; Omar, M.Z.; Abdullah, S. Effects of rheocasting and thixoforming on the microstructure and mechanical properties of A356 aluminium alloy. *J. Teknol.* **2016**, *78*, 107–113.
32. Rooy, E.L.; Linden, J.H.L. *Van ASM Metals Handbook, Vol 02 Properties and Selection: Nonferrous Alloys and Special-Purpose Materials*; ASM International: Geauga, OH, USA, 1990; pp. 3330–3345.
33. Kurz, W.; Fisher, D. *Fundamentals of Solidification*; Trans Tech Publication Ltd.: Aedermannsdorf, Switzerland, 1986; ISBN 0878495223.
34. Flemings, M.C. Behavior of metal alloys in the semisolid state. *Metall. Trans. B* **1991**, *22*, 269–293. [CrossRef]
35. Ogris, E.; Wahlen, A.; Lüchinger, H.; Uggowitzer, P.J. On the silicon spheroidization in Al-Si alloys. *J. Light Met.* **2002**, *2*, 263–269. [CrossRef]
36. Tiryakioğlu, M. Si particle size and aspect ratio distributions in an Al-7%Si-0.6%Mg alloy during solution treatment. *Mater. Sci. Eng. A* **2008**, *473*, 1–6. [CrossRef]
37. Li, B.; Wang, H.; Jie, J.; Wei, Z. Effects of yttrium and heat treatment on the microstructure and tensile properties of Al-7.5Si-0.5Mg alloy. *Mater. Des.* **2011**, *32*, 1617–1622. [CrossRef]
38. Ishak, N.N.M.; Salleh, M.S.; Yahaya, S.H.; Mohamad, E.; Sulaiman, M.A. The Effect of Equal Channel Angular Pressing (ECAP) on the Microstructure and Hardness of A356 Aluminium Alloy. *J. Adv. Manuf. Technol.* **2017**, *11*, 47–58.
39. Cepeda-Jiménez, C.M.; García-Infanta, J.M.; Zhilyaev, A.P.; Ruano, O.A.; Carreño, F. Influence of the supersaturated silicon solid solution concentration on the effectiveness of severe plastic deformation processing in Al–7 wt.% Si casting alloy. *Mater. Sci. Eng. A* **2011**, *528*, 7938–7947. [CrossRef]
40. Ritwik, R.; Rao, A.K.P.; Dhindaw, B.K. Low-convection-cooling slope cast AlSi7Mg alloy: A rheological perspective. *J. Mater. Eng. Perform.* **2013**, *22*, 2487–2492. [CrossRef]
41. Natori, K.; Utsunomiya, H.; Tanaka, T. Improvement in formability of semi-solid cast hypoeutectic Al-Si alloys by equal-channel angular pressing. *J. Mater. Process. Technol.* **2017**, *240*, 240–248. [CrossRef]
42. Mill, E.B.; Sherif, E.M.; Mohammed, J.A.; Abdo, H.S.; Almajid, A.A. Corrosion Behavior in Highly Concentrated Sodium Chloride Solutions of Nanocrystalline Aluminum Processed by High. *Int. J. Electrochem. Sci.* **2016**, *11*, 1355–1369.

43. Nguyen, V.T.; Hussain, Z.; Anasyida, A.S.; Huy, T.D.; Almanar, I.P. Influence of Semi-Solid Casting and Equal Channel Pressing on Microstructure of a Hypoeutectic Al-Si Alloy. *Mater. Sci. Forum* **2015**, *819*, 9–14. [CrossRef]
44. Rollett, A.; Humphreys, F.; Rohrer, G.; Hatherly, M. *Recrystallization and Related Annealing Phenomena*; Elsevier Science Ltd.: Oxford, UK, 2004.
45. Mckenzie, P.W.J.; Lapovok, R.; Estrin, Y. The influence of back pressure on ECAP processed AA 6016: Modeling and experiment. *Acta Mater.* **2007**, *55*, 2985–2993. [CrossRef]
46. Meyers, M.A.; Mishra, A.; Benson, D.J. Mechanical properties of nanocrystalline materials. *Prog. Mater. Sci.* **2006**, *51*, 427–556. [CrossRef]
47. Zhu, M. Effects of T6 heat treatment on the microstructure, tensile properties, and fracture behavior of the modified A356 alloys. *J. Mater.* **2017**, *36*, 243–249. [CrossRef]
48. Möller, H.; Govender, G.; Stumpf, W. Factors influencing tensile mechanical properties of Al-7Si-Mg casting alloys A356/7. In *Light Alloys 2012*; Springer: Berlin, Germany, 2012; pp. 467–471.
49. Goodarzy, M.H.; Arabi, H.; Boutorabi, M.A.; Seyedein, S.H.; Najafabadi, S.H.H. The effects of room temperature ECAP and subsequent aging on mechanical properties of 2024 Al alloy. *J. Alloys Compd.* **2014**, *585*, 753–759. [CrossRef]
50. Kumar, S.R.; Gudimetla, K.; Venkatachalam, P.; Ravisankar, B.; Jayasankar, K. Microstructural and mechanical properties of Al 7075 alloy processed by Equal Channel Angular Pressing. *Mater. Sci. Eng. A* **2012**, *533*, 50–54. [CrossRef]
51. Thuong, N.; Zuhailawati, H.; Seman, A.A.; Huy, T.D.; Dhindaw, B.K. Microstructural evolution and wear characteristics of equal channel angular pressing processed semi-solid-cast hypoeutectic aluminum alloys. *Mater. Des.* **2015**, *67*, 448–456. [CrossRef]
52. Hughes, A.E.; Birbilis, N.; Mol, J.M.C.; Garcia, S.J.; Zhou, X.; Thompson, G.E. Corrosion and principles of protection. In *Recent Trends in Processing and Degradation of Aluminum Alloys*; Intech: Rijeka, Croatia, 1968.
53. Birbilis, N.; Buchheit, R.G. Electrochemical Characteristics of Intermetallic Phases in Aluminum Alloys: An Experimental Survey and Discussion. *J. Electrochem. Soc.* **2005**, *152*, 140–151. [CrossRef]
54. Boag, A.; Hughes, A.E.; Glenn, A.M.; Muster, T.H.; McCulloch, D. Corrosion of AA2024-T3 Part I: Localised corrosion of isolated IM particles. *Corros. Sci.* **2011**, *53*, 17–26. [CrossRef]
55. Balyanov, A.; Kutnyakova, J.; Amirkhanova, N.A.; Stolyarov, V.V.; Valiev, R.Z.; Liao, X.Z.; Zhao, Y.H.; Jiang, Y.B.; Xu, H.F.; Lowe, T.C.; et al. Corrosion resistance of ultra fine-grained Ti. *Scr. Mater.* **2004**, *51*, 225–229. [CrossRef]
56. Di Schino, A.; Kenny, J.M. Effects of the grain size on the corrosion behavior of refined AISI 304 austenitic stainless steels. *J. Mater. Sci. Lett.* **2002**, *21*, 1631–1634. [CrossRef]
57. Yahya, S.; Rahim, A.A. Inhibitive Behaviour of Corrosion of Aluminium Alloy in NaCl by Mangrove Tannin. *Sains Malaysiana* **2011**, *40*, 953–957.
58. Inturi, R.B. Localized Corrosion of Nanocrystalline 304 Type Stainless Steel Films. *Corrosion* **1992**, *48*, 398–403. [CrossRef]
59. Rashida, S.; Islamia, N.; Ariffina, A.K.; Ridhab, M.; Fonnab, S. The effect of immersion time on the corrosion behavior of SUS304 in brine using half-cell potential measurement. *Polarization* **2016**, *5*, 7.
60. Krishna, K.G.; Sivaprasad, K.; Narayanan, T.S.N.S.; Kumar, K.C.H. Localized corrosion of an ultrafine grained Al–4Zn–2Mg alloy produced by cryorolling. *Corros. Sci.* **2012**, *60*, 82–89. [CrossRef]
61. Akiyama, E.; Zhang, Z.; Watanabe, Y. Effects of severe plastic deformation on the corrosion behavior of aluminum alloys. *J. Solid State Electrochem.* **2009**, *13*, 277–282. [CrossRef]
62. Song, D.; Ma, A.; Jiang, J.; Lin, P.; Yang, D. Corrosion behavior of ultra-fine grained industrial pure Al fabricated by ECAP. *Trans. Nonferr. Met. Soc. China Engl. Ed.* **2009**, *19*, 1065–1070. [CrossRef]

metals

MDPI

Article

Superplastic Behavioral Characteristics of Fine-Grained 5A70 Aluminum Alloy

Sheng Li, Zhongguo Huang and Shunyao Jin *

School of Mechanical Engineering, University of Science and Technology Beijing, Beijing 100083, China; lishengsir@163.com (S.L.); jsy_white@126.com (Z.H.)

* Correspondence: jinshunyao@ustb.edu.cn; Tel.: +86-010-6233-2467

Received: 18 December 2018; Accepted: 8 January 2019; Published: 10 January 2019

Abstract: A fine-grained 5A70 alloy sheet was obtained through a combination of rolling and heat treatment, with a total deformation reduction of 90% and an average grain size of 8.48 μm. The alloy was studied at 400, 450, 500, and 550 °C and exhibited excellent elongation-to-failures of 205, 321, 398, and 437% with coefficients for the strain rate sensitivity of 0.42, 0.40, 0.47 and 0.46, respectively. Electron backscatter diffraction (EBSD) results revealed that the massive grain boundaries were high angle boundaries, suggesting that boundary sliding and grain rotation occurred during superplastic deformation. The X-ray diffraction (XRD) and energy dispersive spectrometer (EDS) results indicated that the compositions were the Al_6(MnFe) and Mg-rich phase particles of the deformed 5A70 alloy. In addition, the weakening of the pinning effect led to abnormal grain growth at 500 and 550 °C, resulting in strain hardening. Transmission electron microscopy (TEM) examinations demonstrated that the applied stress at the head of the precipitated particles and/or grain boundaries exceeded the matrix-structure-promoted cavity nucleation. Cavities grew, interlinked, and coalesced, which resulted in crack formation that eventually led to superplastic fractures. Filaments formed at the fracture surfaces because of second phase precipitation at grain boundaries and the formation of Mg-rich oxides.

Keywords: 5A70 aluminum alloy; fine-grained structure; superplastic tension; grain boundary sliding; precipitated phase; strain hardening

1. Introduction

Due to the excellent combination of strength, corrosion resistance, weldability, formability, and low cost, Al-Mg alloys are widely used in the aircraft, shipbuilding, and automotive industries [1]. However, Al-Mg alloys are not heat treatable, so high strengths are achieved through solution hardening using the Mg atoms that are retained in the solid solution, through precipitation hardening from the second phase particles, or through strain hardening effects [2–5]. Accordingly, additional contents of alloying elements, Mn, Ti, and Zr, refine the recrystallized grain size, while the Cu and Mg effectively improve the strength of Al-Mg alloys. The subcrystal and grain sizes of Al-Mg alloys are refined with the increasing Mg concentration. Furthermore, recrystallization can promote the evolution of the grain orientation towards high-angle grain boundaries [6,7]. It is known that the increase in magnesium content (3–8.5%) provides finer grains and improves the elongation-to-failure, δ, in Al-Mg alloys [8–10].

Numerous domestic and literature studies have been conducted concerning different methods used to prepare Al-Mg alloys and investigations on their superplasticity in the preceding decades. The equal channel angular pressing (ECAP) [11–16] and friction stir welding/processing (FSW/FSP) [17–22] technologies were used to prepare the fine/ultrafine grain structure of Al-Mg alloys to investigate their superplasticity. For example, the 5083 alloy was prepared using the FSW method and its superplastic behavior was investigated in the temperature range from 250 to 450 °C [23]. The results revealed that reasonable superplasticity with an elongation-to-failure of 570% was achieved at 300 °C and 8.3×10^{-3} s^{-1}. In addition, the research regarding the superplasticity of Al-Mg alloys with an Mg content lower than 5.5 wt.% and refined structure showed excellent elongation. With the development of the lightweight aviation industry and the requirements of sonic and supersonic aircraft, the superplastic form of aluminum alloy sheet metal can survive severe plastic deformations, reduce the weight, and increase the strength for complex structures. Therefore, high-strength Al-Mg alloys have significant advantages in stamping parts, including the fuselage, stringer, vertical tail skin, etc. [24]. It is well-known that a high Mg content enhances the strength of Al-Mg alloys; however, the precipitation of a large amount of Mg-rich phase particles can lead to strain hardening in superplastic tension. The requirements for military, aluminum-alloy, superplastic sheet products are continually increasing, so the demand for rolling and heat treatment (RHT) methods to prepare superplastic sheets is urgent. In this paper, the RHT method will be used to obtain a fine-grained 5A70 alloy, mainly for large sheet metal parts in industrial applications.

Structural superplasticity is the ability of polycrystalline materials to exhibit high tensile elongations without the formation of a neck prior to fracture because of a high valued strain rate sensitivity, m [25]. Strain rates during superplastic deformation with the fine-grained 5A70 alloy obey the following relationship [26–30]:

$$\dot{\varepsilon} = A\frac{GD_0 b}{kT}\left(\frac{b}{d}\right)^p\left(\frac{\sigma-\sigma_0}{G}\right)^n \exp\left(-\frac{Q}{RT}\right), \quad (1)$$

where, A is a dimensionless material constant; b is the Burgers vector; d is the grain size; σ is the applied stress; σ_0 is the threshold stress; G is the shear modulus; k is Boltzmann's constant; T is the absolute temperature; p is the exponent of the inverse grain size, which ranges from 2 to 3; n is the stress exponent, which is defined as $1/m$; D_0 is a frequency factor; Q is the activation energy; and R is the gas constant. For the majority of superplastic materials, the rate-controlling process of superplastic deformation is grain boundary sliding (GBS) [31]. An analysis of the superplastic tension testing and surface observation showed that lattice diffusion dominated the GBS mechanism in superplastic tension of the 5A70 alloy. This mechanism, through the diffusion activation energy and the strain rate sensitivity, was defined as a constant rate with a temperature dependence of the fine-grained structure. Therefore, this study focused on revealing the superplastic behavior characteristics of the 5A70 alloy with fine-grained structures that depend on temperatures, strain rates, and precipitated phases during superplastic deformation.

2. Materials and Methods

The studied 5A70 alloy (correspond to GB/T 3190-2008 (China)), with a chemical composition of Al-5.72Mg-0.60Mn-0.058Cu-0.20Fe-0.080Si-0.095Zr-0.020Zn-0.043Ti (wt.%), was fabricated using continuous casting. A differential scanning calorimeter (DSC) experiment was conducted using the processed sample on the SDT-Q600 thermo-analytical instrument (TA Instruments Inc., New Castle, PA, USA), and the curve was obtained at a heating rate of 5 °C/min for the 5A70 alloy immediately after solutionizing and natural aging. Meanwhile, the phase-transition temperature of the Al-5.7Mg alloy ranged from 255–575 °C, according to the Al-Mg binary phase diagram shown in Figure 1b [32]. The DSC curve in Figure 1a shows two endothermic peaks. The result of the first endothermic peak clearly indicates the onset of incipient solid solution at 556 °C, and the solid solution temperature of the β phase particles was 581 °C. The second endothermic peak demonstrates that the melting temperature of 5A70 aluminum alloy was 631 °C.

The prepared experimental alloy ingot was homogenized and annealed at 450 °C for 40 h. Then, the ingot alloy was rolled into a billet with a rectangular normal direction plane of 255 mm × 255 mm at 380 °C. A billet with a plate shape (200 mm × 200 mm × 25 mm) was processed from the state of the extruding ingot, and the natural aging treatment was ≥240 h. The size and distribution of the precipitated phases in the smelted and forged processes were controlled to promote nucleation during dynamic recrystallization. The billet was subjected to 14 passes of cold rolling on a 350 mm reversing cold mill. In addition, full recrystallization at 340 °C was performed for 30 min using a GS-2-1200 box-type resistance furnace when the sheet was 10 and 5 mm thick. Ultimately, a 2 mm thick fine-grained 5A70 alloy superplastic sheet was obtained. Further details of the rolling process and the full recrystallization system were reported in a previous work [33].

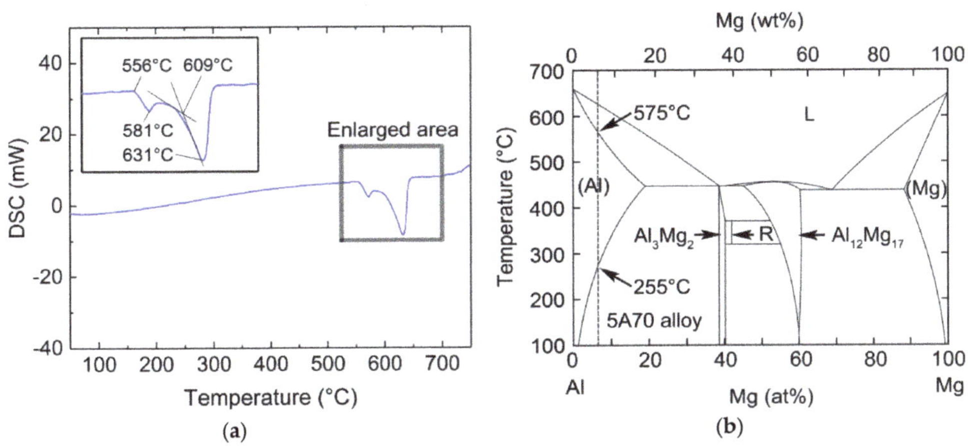

Figure 1. Differential scanning calorimeter (DSC) results of the 2 mm thick fine-grained 5A70 alloy (**a**) and the binary Al-Mg phase diagram (**b**).

Specimens with 8 mm gauge lengths and 4 mm gauge widths were machined along the parallel rolling direction. Superplastic tensile tests were performed on an AG-250KINC Instron machine (Shanghai Gold Casting Instrument Analysis Co., Ltd., Shanghai, China) with a microprocessor control pad in an NV63-CV high temperature furnace. The tests were performed at 400, 450, 500, and 550 °C, and the initial strain rates were 5×10^{-3}, 1×10^{-3}, and 5×10^{-4} s^{-1} in air conditions. In addition, type-K thermocouples were used to detect the furnace temperature, which was controlled within approximately ±2 °C along the entire gauge length during the tests. The specimen was insulated at 340 °C for 10 min for full recrystallization; moreover, the furnace can control the temperature equilibrium during the heating process. The same heating rate, 21 °C/min, was utilized to reach the target temperature within 10 min and held for 2 min for stabilization.

Microstructural characterization and analysis of the 5A70 alloy were carried out using a Jeol-7100 (JEOL Ltd., Tokyo, Japan) transmission electron microscope (TEM), and a 600FEG (FEI Corporation, Hillsboro, OR, USA) scanning electron microscope (SEM). The average grain size was found using the linear intercept method with the OIM software (6.2.0 x86 version, EDAX Inc., Draper, UT, USA). The boundary orientation was measured by utilizing pixel-to-pixel measurements. To carry out the structural characterization in the SEM, specimens were cut into 5 mm sections near the superplastic fracture surface.

3. Results

3.1. Initial Microstructures

The 2 mm thick, deformed, superplastic sheet was fully recrystallized at 340 °C for 30 min, while the total deformation was 90% (20–2 mm). The electron backscatter diffraction (EBSD) map of the studied alloy based on the inverse pole figure coloring (insert) is shown in Figure 2. The average grain size of the fully recrystallized structure was 7.80 μm, the fraction of the high angle grain boundaries (HAGBs, $\theta \geq 15°$) was 0.96, and the average orientation angle was 38.0° in the rolling direction (RD) plane, as shown in Figure 2a,d,e. Similarly, Figure 2b,d,e show the recrystallization structure along the normal direction (ND) plane. The average grain size was 8.4 μm, the HAGBs were 0.51, and the average of the orientation angle was 19.4°. Figure 2c,d,e show the recrystallization structure along the transverse direction (TD) plane. The average grain size was 9.3 μm, the HAGBs were 0.85, and the average of the orientation angle was 35°. The GBS can frequently take place during superplastic deformation because of the fine-grained structure and high fraction of the HAGBs in the initial microstructures.

Grain fragmentation followed by rolling deformation of new grains rapidly occurred in the original grain interiors after full recrystallization, as shown in Figure 2a–c. The new grains exhibited ultrafine equiaxed structures along the original boundaries. The grain size distributions were uniform over all planes, and most of the grain sizes ranged from 5 to 10 μm. However, the low angle grain boundaries (LAGBs, θ = 2–15°) were as high as 57.4% in the ND plane whenever there was only 8.3 and 14.7% in the RD and TD planes, respectively. In summary, the larger the grain orientation angle and the higher the grain boundary index, the smaller the fine-grained size (average grain size = $\sqrt[3]{7.8 \times 8.4 \times 9.3} = 8.48$ μm).

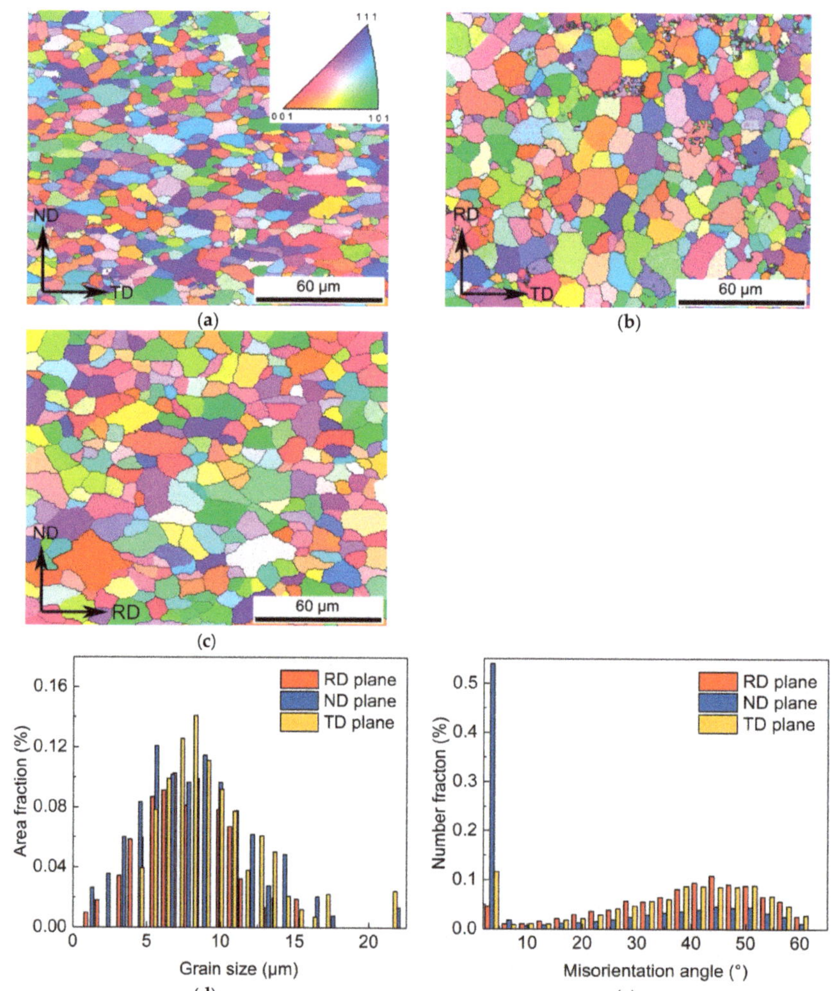

Figure 2. EBSD results of the grain size in the RD plane (**a**), ND plane (**b**), and TD plane (**c**); the grain size distribution (**d**); and the grain orientation angle (**e**) of the 2 mm thick fine-grained 5A70 alloy.

3.2. Superplastic Deformation Behaviors

3.2.1. Elongation-to-Failure

Superplastic tensile tests focused on investigating the deformation mechanism at different temperatures, strain rates, and influences of the precipitated phases. The superplastic elongation-to-failure, δ, results for the 5A70 alloy are shown in Figure 3.

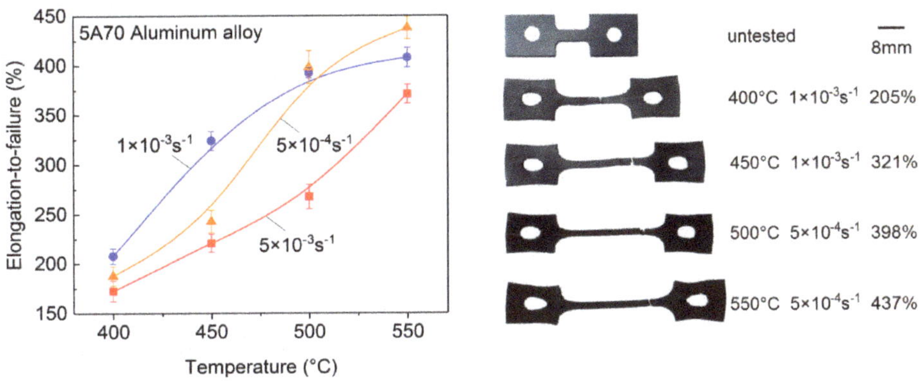

(a) (b)

Figure 3. Superplastic elongation-to-failure results (**a**) and the specimens with excellent δ values at different temperatures, 400–550 °C, and different initial strain rates, 5×10^{-4}–5×10^{-3} s^{-1} (**b**).

Figure 3a shows that the elongation-to-failure increased gradually with increasing temperatures and decreasing initial strain rates. Figure 3b shows that the samples were pulled to failure at different conditions, where the top image is the untested specimen. The 5A70 alloy exhibited reasonable δ values at 400 and 450 °C, with a moderate strain rate. However, excellent superplastic elongation-to-failures gave low strain rates at 500 and 550 °C, and the maximum δ value was 437% at 550 °C with 5×10^{-4} s^{-1}. In addition, necking was not visible in the superplastic fracture surface and a stable superplasticity elongation was exhibited simultaneously.

3.2.2. True Strain-True Stress

Figure 4 shows the superplastic tensile results for the true strain-true stress of the fine-grained 5A70 alloy at 400–550 °C. At a constant temperature, the strain rate enhancement phenomenon of the materials was consistent with the general law of superplastic elongation characteristics and increased strain rate. The peak applied stress can be observed without the steady-state flow stage, which was the typical variation of the true stress-true strain behavior of the Al-Mg alloys [34–36]. The strain hardening was attributed to dislocation sliding/climbing. In addition, the dislocation density changed non-monotonically with the stable grain structure during the initial stages of superplastic deformation. A high dislocation density at the initial deformation for T = 400 °C with a high strain rate was caused by grain adaptation, i.e., the dislocation density increased rapidly and plugged in the grain, forming dislocation walls/cells and leading to an increased true stress. However, with the accumulation of deformation, the grain rotation occurred under shear stresses, and dislocations were absorbed by the grain boundaries, which led to the true stress remaining stable over a short period of time [37]. This is the reason that the true stress presented a step-up state. As the temperatures increased and the superplastic deformation accumulated, the true stress increased and the strain hardening gradually enhanced. At 500 and 550 °C with 5×10^{-4} s^{-1}, superplastic δ values achieved 398 and 437% due to the strong reduction of the unstable superplastic flow. However, the effect of diffusion creep was continuously strengthened as the deformation accumulated. The strong pinning effect declined because of the dissolution of the precipitated second phase particles, and the decreasing of the pinning effect led to abnormal grain growth during superplastic tensile. In addition, the completion of the grain rotation at the final tensile stage, which cannot balance the shear stress caused by the true stress, continually increased, as shown in Figure 4b.

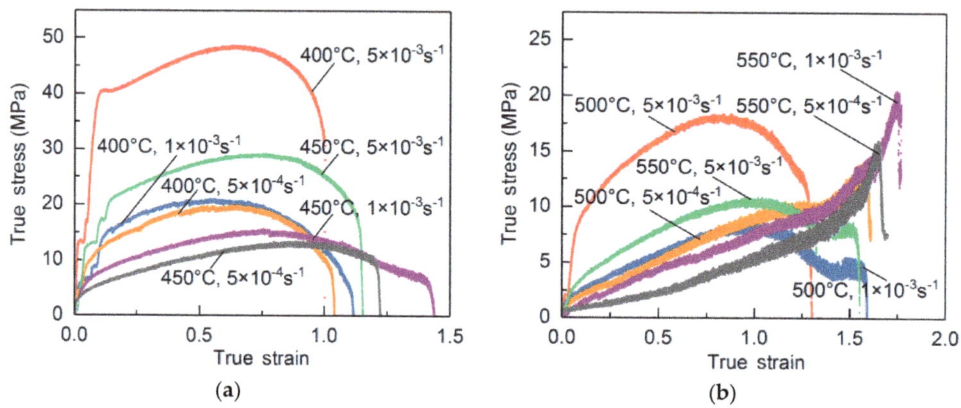

Figure 4. True strain-true stress results for the fine-grained 5A70 alloy at different temperatures: 400–450 °C (**a**) and 500–550 °C (**b**) with strain rates: 5×10^{-3}–5×10^{-4} s^{-1}.

3.2.3. Strain Rate Sensitivity

The coefficient of strain rate sensitivity, m, indicates the capability of an alloy to resist necking spread and symbolizes the values of the superplastic elongation-to-failure. The m value was derived as follows:

$$m = \frac{\partial \ln \sigma}{\partial \ln \dot{\varepsilon}}, \tag{2}$$

The applied stress and strain rate data were plotted in the ln(σ)-ln($\dot{\varepsilon}$) coordinate system. The coefficients of strain rate sensitivity at different temperatures and strain rates were obtained at ε = 0.6 and are shown in Figure 5a. The variation of the m values of the 5A70 alloy with the true strain in the superplastic deformation are shown in Figure 5b.

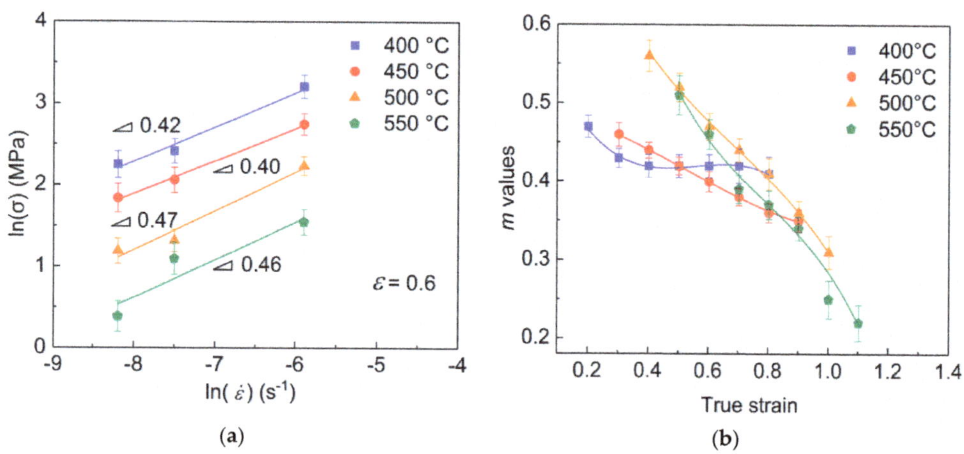

Figure 5. Coefficients of the strain rate sensitivity, m, at ε = 0.6 (**a**) and the m values during superplastic deformation at 400–550 °C (**b**).

It has been well-documented that the solute drag creep is the dominant deformation mechanism in the coarse-grained/fine-grained Al-Mg alloys at low temperatures and high strain rates [38]. The solute drag creep was previously characterized for $m \leq 0.33$ [39,40]. However, the apparent m values in the present study were generally higher than 0.33 and close to 0.5. Thus, the solute drag creep might make a relatively limited contribution to the observed superplasticity. In addition, the corresponding stress exponent, n, values ranged from 2.13–2.50, which were close to the typical model for the superplasticity stress exponent (n = 2) [41]. Under the experimental temperatures and strain rates, the m values declined with the accumulation of the deformation strain. Therefore, at 400 and 450 °C, necking was prevented due to an extensive strain-hardening rate and the 5A70 alloy exhibited a moderate tensile ductility with $m \geq 0.33$. At 500 and 550 °C, both strain hardening and the strain-hardening rates associated with decreased necking had low strain rates and high δ values.

3.2.4. Activation Energy for Flow

The activation energy for flow in superplastic deformation was used to assess the degree of difficulty of the material thermal deformation or for the rearrangement of atoms. The higher the dislocation density of the deformation alloy, the greater the activation energy. The activation energy, Q, was obtained by taking the logarithm of Equation (1) under constant strain rates. The apparent diffusion activation energy of Q was derived as follows:

$$Q = nR\frac{\partial \ln(\sigma - \sigma_0)}{\partial(1/T)}, \quad (3)$$

The variation in the logarithmic stress as a function of reciprocal temperatures illustrates that a low strain rate appeared to be more sensitive to temperatures. The activation energy for flow, Q, ranged from 135 to 139 kJ/mol and is illustrated in Figure 6. In addition, the lattice diffusion activation energy of Q_L = 143.4 kJ/mol [42] and the grain boundary diffusion activation energy of Q_{gb} = 84.0 kJ/mol [43] for pure aluminum demonstrated that lattice diffusion dominated the GBS diffusion processes. The activation energy for flow, Q, was 139 kJ/mol, and the maximum δ of 437% was obtained at 550 °C and 5×10^{-4} s^{-1}.

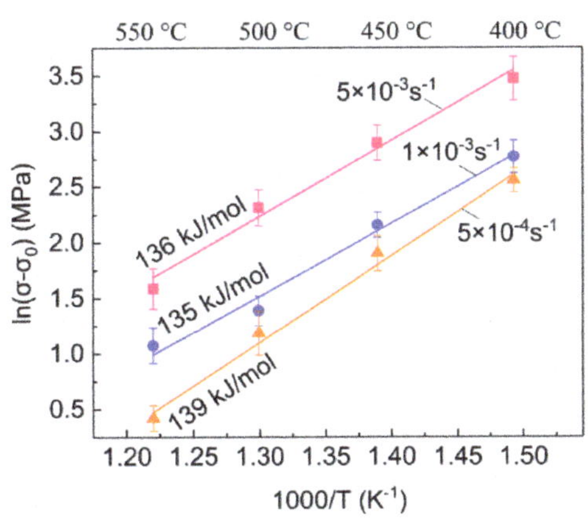

Figure 6. Logarithmic stress as a function of reciprocal temperatures and the apparent activation energy Q of the fine-grained 5A70 alloy.

3.3. Superplastic Fracture Surface Morphologies

The superplastic fracture specimens were naturally aged, and the SEM results of the fracture surface morphologies at different temperatures and strain rates are shown in Figure 7. Fracture surfaces that formed under the GBS exhibited the characteristics of ductile failure at grain boundaries, and a large density of submicrometer filaments was evident on the fracture surfaces. Nevertheless, these filaments protruded from the fracture surface along the tensile direction. Similar sub-micrometer filaments were previously observed on the fracture surfaces of deformed superplastic alloys AA5083 [38] and AA7475 [44], as well as for aluminum-matrix composites [45].

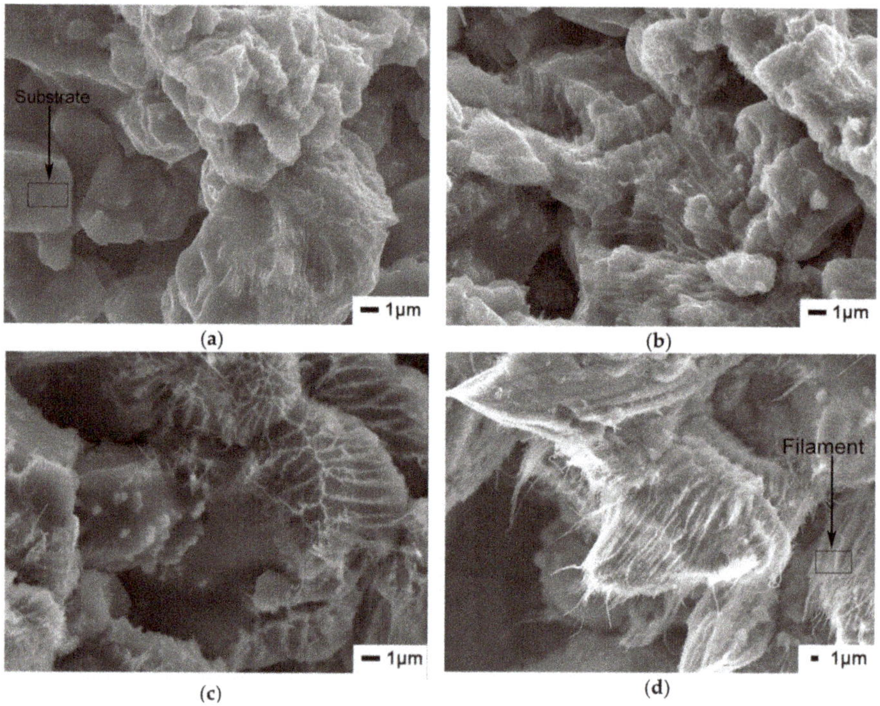

Figure 7. SEM micrograph of the superplastic fracture surface morphologies at T = 400 °C, $\dot{\varepsilon} = 1 \times 10^{-3}$ s^{-1} (**a**), T = 450 °C, $\dot{\varepsilon} = 1 \times 10^{-3}$ s^{-1} (**b**), T = 500 °C, $\dot{\varepsilon} = 5 \times 10^{-4}$ s^{-1} (**c**), and T = 550 °C, $\dot{\varepsilon} = 5 \times 10^{-4}$ s^{-1} (**d**).

A variety of explanations have been reported for the origins of these sub-micrometer filaments. Superplastic tensile tests were performed below the temperature of the initial phase inversion (T < 556 °C). Therefore, the formation of filaments at 400 °C was actually due to sliding transitions from solute drag creeping to high angle grain boundaries. However, this cannot be from the incipient melting or the formation of a glassy phase at grain boundaries [46]. Figure 8, however, provides apparent evidence from which higher concentrations of magnesium and oxygen were observed on the fracture surfaces, indicating the formation of an Mg-rich oxide. Oxides were ridged on early exposed surface regions, as shown in Figure 8, with a composition similar to the filaments on the fracture surface. Ritchie et al. [47] reported that the Mg-rich oxides MgO and $MgAl_2O_4$ were preferentially formed on Al-Mg alloys at 500 °C. Chang et al. [48] confirmed that filaments were observed at the fracture surfaces when the specimen was tested in air rather than under vacuum conditions.

Previous works observed that filament formation occurred in AA5083 within the cracks in the tribolayers that correspond to locations above the grain boundaries of the near-surface subjected to the GBS at T > 300 °C [49,50]. Consequently, the Mg_5Si_6 phase precipitated at the grain boundaries. This was the reason for the increased Si content from 2.95 (in the substrate) to 11.69 wt.% (in the filaments). In addition, the Mg_5Si_6 phase particles that precipitated along the grain boundaries and dominated the filament formation were the result of grain boundary sliding during superplastic deformation. Therefore, this is a reasonable explanation for the increased Si content in the filaments.

These results suggest that the formation of sub-micrometer filaments of the fine-grained 5A70 alloy during superplastic deformation occurred from the growth of Mg-rich oxides and precipitated second phases at the grain boundaries.

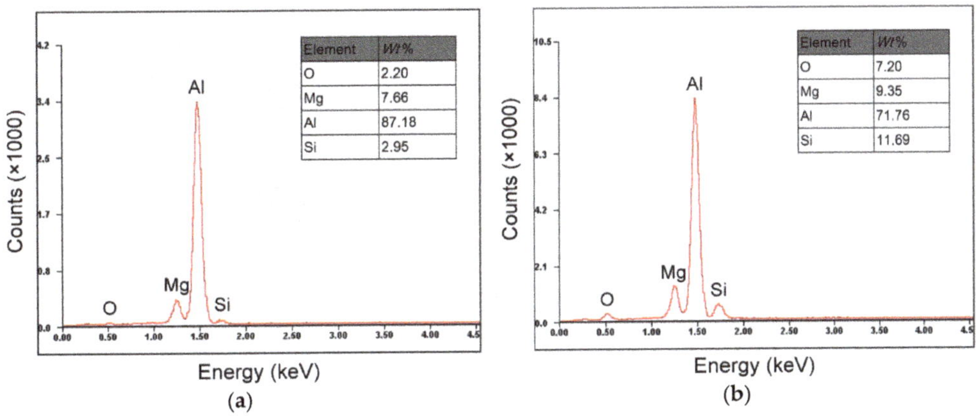

Figure 8. EDS analysis for the superplastic fracture surface of the substrate (**a**) and the filaments (**b**) of the fine-grained 5A70 alloy.

4. Discussion

4.1. Influence of Plastic Flow during Superplastic Deformation

It is known that strain hardening and/or the strain-hardening rate can provide the plastic stability during superplastic tensile, defined by:

$$\sigma = K_1 \dot{\varepsilon}^m \varepsilon^{n_1}, \tag{4}$$

where, K_1 is a constant, and n_1 is the coefficient of strain hardening. A logarithmic analysis of Equation (4) showed that $n_1 = dln\sigma/dln\varepsilon$ was calculated by taking the true stress and strain rates of the true strain, ε = 0.3–0.7, in the superplastic ductility, and the resulting n_1 values are shown in Table 1.

Table 1. Coefficients of strain hardening at different temperatures and strain rates.

Temperature T/°C	Strain Rate $\dot{\varepsilon}/s^{-1}$	Strain Hardening Coefficient/n_1	Standard Deviation/%	Temperature T/°C	Strain Rate $\dot{\varepsilon}/s^{-1}$	Strain Hardening Coefficient/n_1	Standard Deviation/%
400	5×10^{-3}	0.14	0.75	500	5×10^{-3}	0.34	1.38
	1×10^{-3}	0.16	0.72		1×10^{-3}	0.61	3.86
	5×10^{-4}	0.24	0.84		5×10^{-4}	0.79	2.23
450	5×10^{-3}	0.23	1.26	550	5×10^{-3}	0.55	2.20
	1×10^{-3}	0.38	1.88		1×10^{-3}	0.69	3.98
	5×10^{-4}	0.46	1.54		5×10^{-4}	0.75	2.78

There is no strain hardening in traditional superplastic materials [51]. In contrast, the 5A70 alloy showed extensive initial strain hardening, similar to other aluminum alloys with fine-grained structures [52]. In particular, the strain hardening at high temperatures and low strain rates is shown in Figure 4b, which significantly contributed to the elongation-to-failure of the superplastic tensile because the coefficient of strain rate sensitivity m was greater than 0.33. The increased dislocation density in the crystal lattice and the distortion of the grain structure were the main reasons for the strain hardening [53]. These m values were not sufficient to stop unstable plastic flow because of the pure strain rate hardening. However, the coefficient of strain hardening was relatively high (Table 1), and the strain hardening attributed to grain coarsening provided unstable plastic flow. Since the true strain rate declined with the accumulated strain, the m values provided a uniform deformation that was visible within the gauge length up to the fracture during the strain-softening stage (Figure 3). The initial strain hardening in the superplastic flow stage provided stability of the plastic flow at the initial stage, with high *m* values.

The temperature-controlled shear modulus G (MPa) of pure aluminum, $G = (3.022 \times 10^4) - 16\,T$, was used [54]. Equation (1) with $p = 2$ and $n = 2$ is typically used to describe the superplastic flow of aluminum alloys with grains in the range of 1–10 μm [55]:

$$\sigma = \sqrt{\frac{GkT}{AD_{gb}}\frac{d}{b^{3/2}}\dot{\varepsilon}^{1/2}}, \tag{5}$$

A plot of σ against $\dot{\varepsilon}^{1/2}$ using double linear scales was adopted to determine the threshold stress. Using the superplastic data gave the best linear fit for the assumed stress exponents for all the investigated temperatures. Therefore, all the values for the threshold stress were estimated by extrapolating the data to zero strain rates using rectilinear regression, as illustrated in Figure 9. The calculated threshold stresses were highly dependent on the deformation temperatures, as summarized in Table 2. The results demonstrated that the threshold stress was associated with a high density of dispersed particles, which impeded the movement of dislocations and grain boundaries during superplastic deformation. For the studied fine-grained 5A70 alloy, the *m* values tend to increase with increasing strain rate over temperatures ranging from 400 to 550 °C. Similarly, the decreased *m* value of the strain accumulation and the reduced superplastic flow stress led to the superplastic fracture.

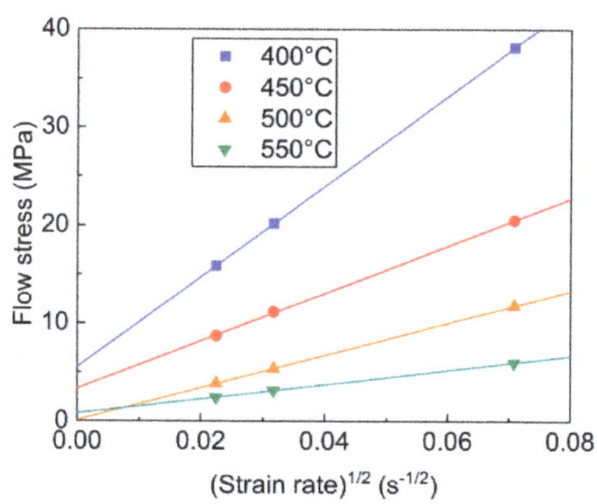

Figure 9. Variation of the flow stress as a function of $\dot{\varepsilon}^{1/2}$ for the fine-grained 5A70 alloy subjected to RHT.

Table 2. Threshold stress of RHT 5A70 alloy at the studied temperatures.

Temperature (°C)	400	450	500	550
Threshold stress (MPa)	5.56	3.35	0.15	0.83

The GBS was closely related to the microstructure recrystallization. In this case, the fine-grained structure and the aberration transformation of the lattice distortion produced by the dislocation slipping/climbing allowed the strain hardening rate to be eliminated during superplastic deformation. In addition, the full recrystallization of the deformed structure evidently does not attenuate the strain hardening. In contrast to the adopted true strain, the *m* values ranged from 0.56 to 0.38, as illustrated in Figure 5b. Nevertheless, the strain hardening occurred in the last stage of the formation, as shown in Figure 4b, and the strain hardening coefficients were 0.61 and 0.55, as shown in Table 1. Moreover, the strain hardening intensified during the superplastic deformation and corresponded to the coefficients of strain hardening of n_1 = 0.79, 0.89 and 0.75. Meanwhile, it was not sufficient to compensate for extensive strain hardening at $m \leq 0.33$, and the true stress under superplastic tensile began to show a general increase. However, the unstable superplastic flow occurred up to $n_1 \leq 0.50$ without strain hardening, as shown in Figure 4a. The superplastic fracture for the fine-grained 5A70 alloy was accompanied by a sharp decline in the deformation and unsteady flow stresses for the strain hardening and strain rate sensitivity at high temperatures.

The influence of temperature and grain size on the diffusion coefficient during the superplastic deformation process in the typical superplastic flow theory of fine-grained alloys was proven through experiments. The effect of the diffusion process was related to the effective diffusion coefficient D_{eff}, which included the lattice diffusion coefficient, D_L, and the grain boundary diffusion coefficient, D_{gb}, whose effective diffusion coefficient is defined as follows [56]:

$$D_{eff} = D_L + x\frac{\pi w}{d}D_{gb}, \tag{6}$$

where x is a constant equal to 1.7 × 10^{-2} and w is the grain boundary width (w = 2b and b = 2.863 × 10^{-10} m) [30]. Equation (6) indicates that the two diffusion paths were independent, and both D_{gb} and D_L contribute simultaneously to the superplastic deformation. As is well-known, the D_L of pure aluminum is D_L (m^2/s) = 1.86 × 10^{-4} exp(−143400/RT), and the D_{gb} of pure aluminum is D_{gb} (m^2/s) = 10^{-4} exp(−84000/RT) [41]. In this study, at 400 and 550 °C, using the Equation: $\varphi = x(\pi w/d)(D_{gb}/D_L)$, φ = 0.022–0.085 < 1 was obtained. Therefore, the dominant diffusion mechanism at the temperature range from 400 to 550 °C was lattice diffusion.

4.2. Effect of Temperature on the Grain Growth and Superplastic Behavior

Equation (1) shows that with a constant strain rate, the superplastic elongation temperature is inversely proportional to the n-th power of the true strain. That is, the higher the temperature, the longer the true strain of the superplastic δ value. Meanwhile, the true stress evidently declines. Figure 10 shows the EBSD analyses of the 5A70 alloy deformed at different temperatures for $\dot{\varepsilon}$= 1 × 10^{-3} s^{-1} after the superplastic fracture. The color of each grain was coded by its crystal orientation based on the [001] inverse pole figure in Figure 10a. The aggregation of a large number of ultrafine grains was near the small cavities in the tensile specimen. In addition, the cavity interlinkage and coalescence were precisely identified via the supporting microscopy evidence. Nevertheless, new ultrafine grains occurred in the limited region and were generated near the initial deformed grains, indicating that dynamic recrystallization occurred. Compared with the microstructures of the sample before deformation (Figure 2b), it can be inferred from Figure 10a–d that the grains were gradually elongated in the tensile direction as the temperature increased.

Figure 10a–d show that the microstructure mainly consists of grain sizes that were larger than 10 μm. Dynamic recrystallization occurred during superplastic deformation, which generated the recrystallized grains. At 400 °C and 1 × 10^3 s^{-1}, dynamic recrystallization occurred without obvious grain growth, giving a grain size of 9.60 μm. The grain boundary character distribution data and average grain sizes of the samples in Figure 10 are listed in Table 3. Compared with Figure 2b, it can be observed that the recrystallized grains gradually coarsen when increasing the deformation temperatures or increasing the deformation degree. However, the final grain structure had an average recrystallized grain size of less than 15 μm for 400–500 °C in this work, revealing that the 5A70 alloy had a strong ability to inhibit the grain growth during superplastic deformation. The proportion of the LAGBs fraction at 400, 450, and 500 °C gradually increased from 11.4 to 16.4 and 22.8%, but decreased to 18.5% at 550 °C. In contrast, the corresponding grain boundary angle decreased initially and then increased, which was due to the distorted grain structure at 550 °C. These fractions indicated that most boundaries were at high-angles, which can indirectly prove the occurrence of grain boundary or interphase boundary sliding. Nevertheless, the abnormal grain growth led to an increased LAGBs and grain misorientation angle, as shown in Figure 10d,h. The grain boundary sliding along the

[110] crystal direction and the orientation grain boundary angle account for more than 50% during the superplastic deformation. Sakai et al. [57] found that the accompanying grain refinement and grain boundary rotation from the large flow softening took place due to the operation of GBS at low strain rates. However, it is known that the glide plane of the face-centered cubic structure in the 5A70 alloy is {111}, and the grain boundary slip direction is <110> [58]. This demonstrates that the sliding mainly took place in the slip direction to balance the grain-to-grain deformation and the reaction stress during superplastic tensile. It is noted that there was positive evidence for the grain rotation and grain boundary sliding for the 5A70 alloy during superplastic deformation.

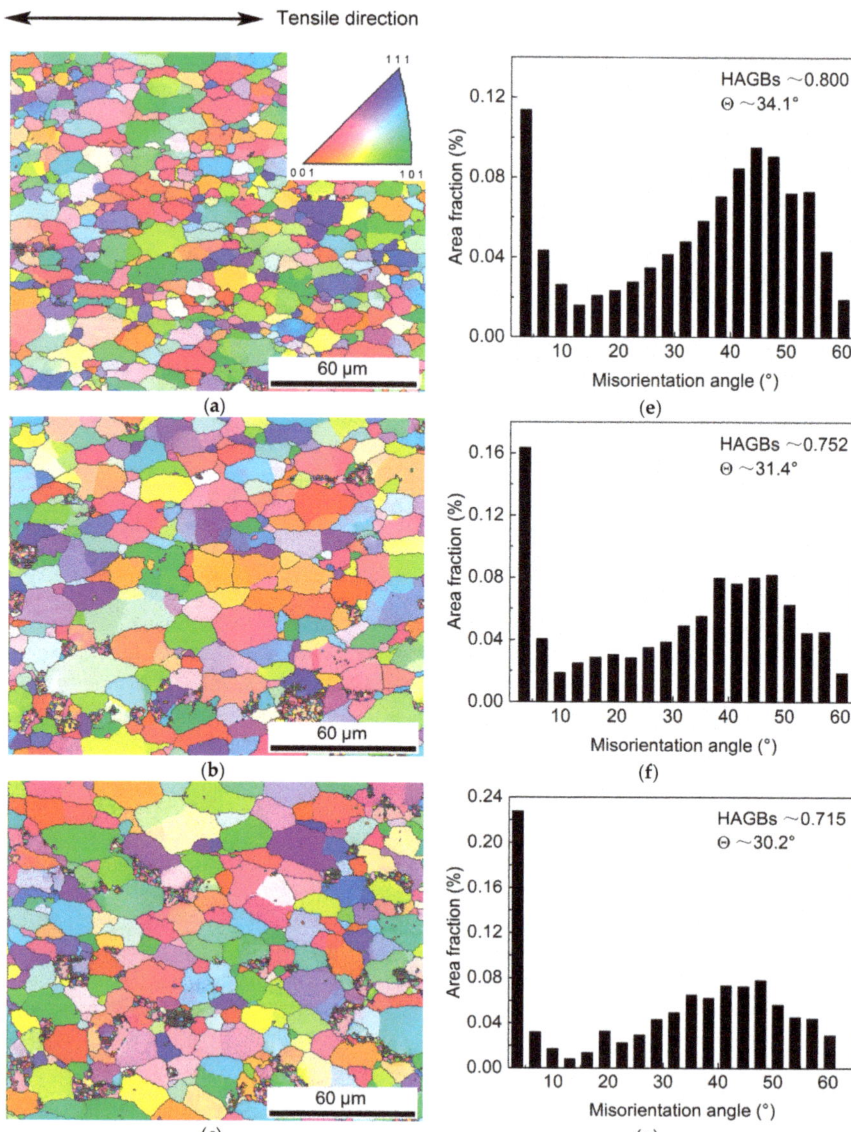

Figure 10. *Cont.*

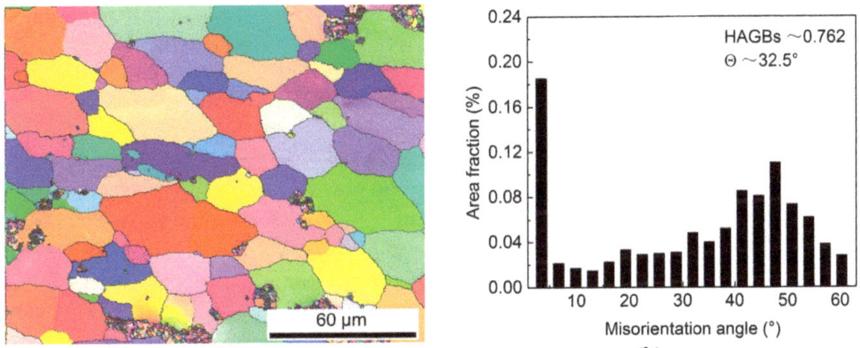

(d) (h)

Figure 10. Influence of the temperature on the grain growth in the deformation process at 400 °C with $\dot{\varepsilon} = 1 \times 10^{-3}$ s^{-1}, where the grain and orientation angle are shown in (**a**,**e**), as well as 450 °C in (**b**,**f**), 500 °C in (**c**,**g**), and 550 °C in (**d**,**h**).

The grain growth of the fine-grained structure was based on the initial grain size during the fully dynamic recrystallization, which includes the static growth of the grain over a certain period of time, and the grain size increased under superplastic tensile. Therefore, the total grain growth during superplasticity is defined as $dD = (\partial D_t/\partial t)dt + (\partial D_\varepsilon/\partial \varepsilon)d\varepsilon$. Sato et al. [59] proposed the deformation induced grain growth (DIGG) model as: $\ln(D/D_0) = \alpha\varepsilon$. Based on the grain growth model introduced by Cao [60], the superplastic deformation grain growth model at a constant rate is shown as follows [61]:

$$d = \left[{d_0}^q \exp(\alpha q \varepsilon) + (K/(\alpha q \dot{\varepsilon}_0)) \exp \varepsilon (\exp(\alpha q \varepsilon) - 1)\right]^{1/q}, \quad (7)$$

where, d is the grain size at the tensile time t, d_0 is the initial grain size, α is the grain growth exponent, q is the growth exponent, ε is the true strain, K is the grain growth rate constant, and $\dot{\varepsilon}_0$ is the initial strain rate. Then, the relation: $\phi = wD_{gb}/(D_0 D_L)$ is used, where w is the grain boundary width. Malopheyev et al. [62] obtained the value for the diffusion coefficient: at 400–550 °C, the calculated available ϕ = 0.02–0.13 < 1. Therefore, the lattice diffusion growth dominated the grain growth mechanism, where q = 3 [63]. The growth rate factor, K, was obtained with the temperature changes and strain rates shown in Table 3.

Table 3. Analysis of the grain growth for the superplastic fracture surfaces.

Temperature T/°C	Strain Rate $\dot{\varepsilon}$/s^{-1}	True Strain ε/MPa	Experimental Results d/μm	Proportional Constant α	Growth Rate Factor ($\times 10^{-21}$) K/m^3 s^{-1}
400	1×10^{-3}	1.13	9.60	0.18	2.32
450		1.45	11.78	0.28	2.53
500		1.60	13.32	0.33	2.58
550		1.61	21.16	0.61	8.94
500	5×10^{-4}	1.76	14.35	0.34	3.42
550		1.71	24.26	0.66	3.47

For temperatures of 400–500 °C, growth of the lattice diffusion promoted the grain structure distortion energy for the grain growth during superplastic deformation. Namely, grain growth along the lattice diffusion of the polymerization growth factors was α = 0.18–0.34, and the grain growth rate was $K = 2.32 \times 10^{-21}$–3.42×10^{-21} m^3/s. This is because the dynamic recrystallization of the deformed structure is a function of the pinning effect of the precipitated particles. Moreover, the grain growth occurred towards the tensile direction as the tensile deformation was accumulated. At 550 °C, the lattice diffusion of the polymerization growth factors was α = 0.61 and 0.66. The abnormal grain growth was due to the weakening of the pinning effect from the dissolution of the precipitated particles, which were unable to effectively inhibit the grain growth during dynamic recrystallization [64]. In addition, the deformed distortion structure had a pronounced orientation along the tensile direction. Therefore, the grain growth exponent, α, indicates the influence of the dynamic grain growth for the 5A70 alloy superplasticity. Furthermore, the superplastic tensile of the 5A70 alloy exhibited a strong temperature dependence. This clearly clarified that the strain hardening was due to the distortion of the grain structure during dynamic recrystallization.

The lattice diffusion dominated the mechanism of the GBS-induced grain to rotate the slip surface under an applied stress and balance the stress tensor to maintain the superplastic flow. At T = 500 °C, the decreased precipitated particles played a significant role in promoting the recrystallized grain growth, which resulted in abnormal grain growth.

4.3. *Influence of Mg-Rich Phase Particles on Superplastic Tensile and Fracture Process*

At 500 °C and 1×10^3 s^{-1}, the components of the precipitated phase were analyzed using X-ray diffraction (XRD) and energy dispersive spectrometry (EDS), as shown in Figure 11. Full recrystallization of the deformed fine-grained 5A70 alloy structure after cold rolling was obtained using heat treatment. Figure 11a shows the dispersive fine precipitates in the recrystallized structure of the rolling surface. In addition, Figure 11b shows the results of the particle determination for the superplastic fracture specimens. There was a transition between the metastable β-Al_2Mg phase and the β-Al_3Mg_2 phase with a hexagonal structure in the dynamic stretching process, as shown in Figure 11c [65]. Figure 11b,d indicate that the Al_6(MnFe) phase disperses after all the heat treatments [30,66], and the Mg_5Si_6 phase precipitated during the superplastic deformation. Similar dispersed phase particles were previously observed in the aluminum-magnesium-silicon alloys [67,68].

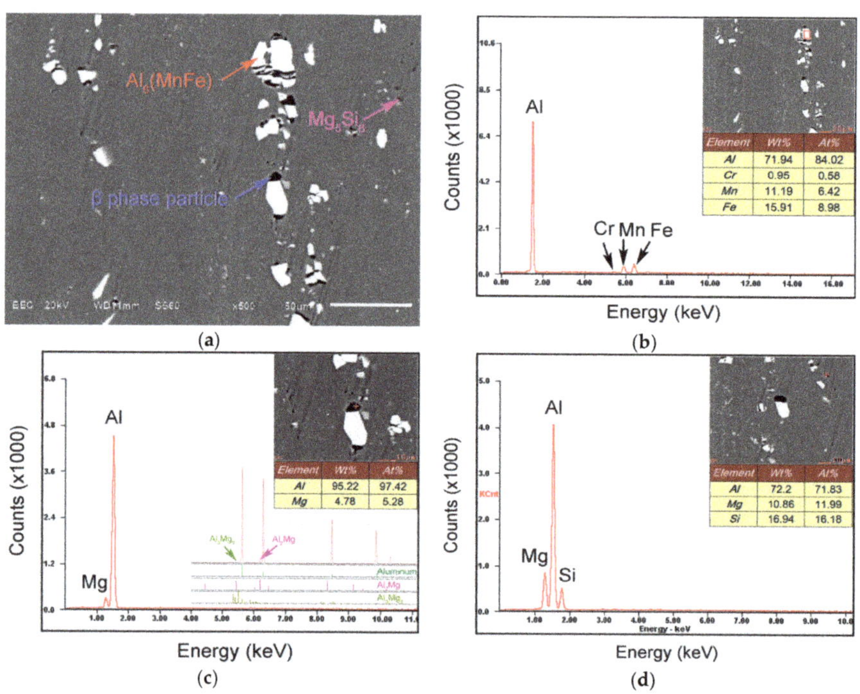

Figure 11. Formation and composition of the dispersed phase particle morphology at full recrystallization (**a**), and the EDS/X-ray diffraction results of Al_6(Mn,Fe) phase (**b**), β phase (**c**), and Mg_5Si_6 phase (**d**).

Precipitation hardening, where small particles inhibit the movement of dislocations to strengthen aluminum alloys, was used to improve the mechanical strength of Al-Mg alloys. The composition and structure of the Mg_5Si_6 phase in aluminum alloys were determined, which occurred as precipitates, and were associated with a particularly strong increase in the mechanical strength. This was due to the magnesium content increasing to 5.72 wt.%, while the addition of Mn, Fe, Cr, and Si improved the nucleation of the second phase particles [69,70]. However, the dynamic recrystallization was intensely impacted by the increased temperatures, and the dispersed distribution of the precipitated phases effectively impeded the grain growth and promoted the equiaxial transformation of the fine-grained structure, as shown in Figure 10. At 550 °C and 1×10^{-3} s^{-1}, the inhibiting effect on the grain growth during tensile of the dynamic recrystallization decreased significantly when the grain size was 21.16 μm, as shown in Figure 10d. The abnormal grain growth resulted in an increase of the true stress under strain hardening (Figure 4b). The pinning effect of the precipitate was the intrinsic mechanism for the change in the true stress-true strain during the superplastic deformation and had a positive effect on the cavity nucleation and growth.

At 500 °C and 1×10^{-3} s^{-1}, when the accumulated applied stress reached a maximum during the superplastic tensile, it was found that sliding of the grain boundary under the shear stress caused dislocations to pin up at the head of the phase particles, as shown in Figure 12a, and the cross grain boundaries slid and climbed to form a sub-grain boundary, as shown in Figure 12b. The TEM results suggest that the dislocation density in the fine-grained structure of the 5A70 alloy during superplastic deformation was ~5×10^{-14} m^2. In addition, the dislocations gradually moved towards and were absorbed by the sub-grain boundaries during superplastic deformation. When the pile-up stress, σ_p, exceeded the theoretical decohesion strength of the Al-matrix/second phase particle interface boundary), the cavity began to nucleate [36]. At 500 °C and 1×10^{-3} s^{-1}, the maximum applied stress (σ_{max} = 3.75 MPa) is substituted for the typical product stress σ_p = 7.87 MPa [71]. Since the plugging stress threshold was more than twice the applied stress during superplastic tensile, the Al-matrix and the strengthening phase particles were easily separated and promoted cavity nucleation.

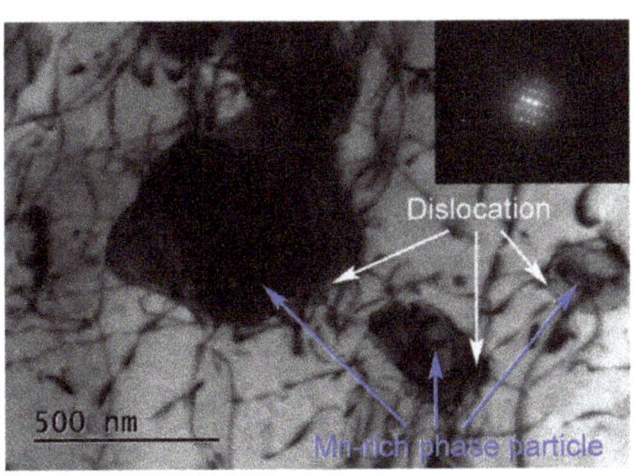

(a)

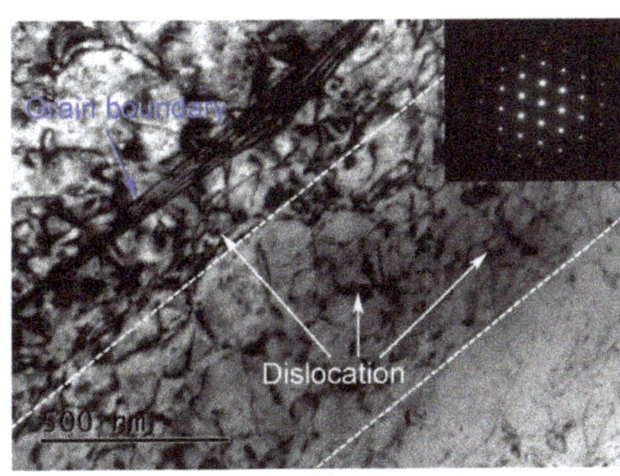

(b)

Figure 12. Location of dislocations, precipitated particles (**a**), and grain boundaries (**b**) in TEM images at T = 500 °C and $\dot{\varepsilon} = 1 \times 10^{-3}$ s^{-1} with ε = 0.65.

The cavity nucleation and growth mechanism of the superplastic tensile was the cavity growth of stress promoting the diffusion of the small cavity along the grain boundary, including the superplastic diffusion growth, and the plastic-controlled growth caused by the plastic deformation of the recrystallized grain around the cavity [72]. The pinning effect of the precipitated phase enhanced the cavity nucleation and the chemical potential between the forceful grain boundary atoms and the free surface of the cavity at 550 °C [73].

Cavity expansion mainly occurred during the growth of the diffusion and the superplastic diffusion because of the spreading of voids into the nearby cavity since the cavity radius was less than the grain size. In addition, plastic-controlled growth dominated the cavity interlinkage and coalescence process, which eventually led to superplastic fracturing [74]. The expansion of the cracks was the fundamental reason for the transient instability and superplastic fracture. The behavior of cavities nearby the fracture surface was studied, as shown in Figure 13, which illustrates the fracture morphology of the superplastic tensile specimens.

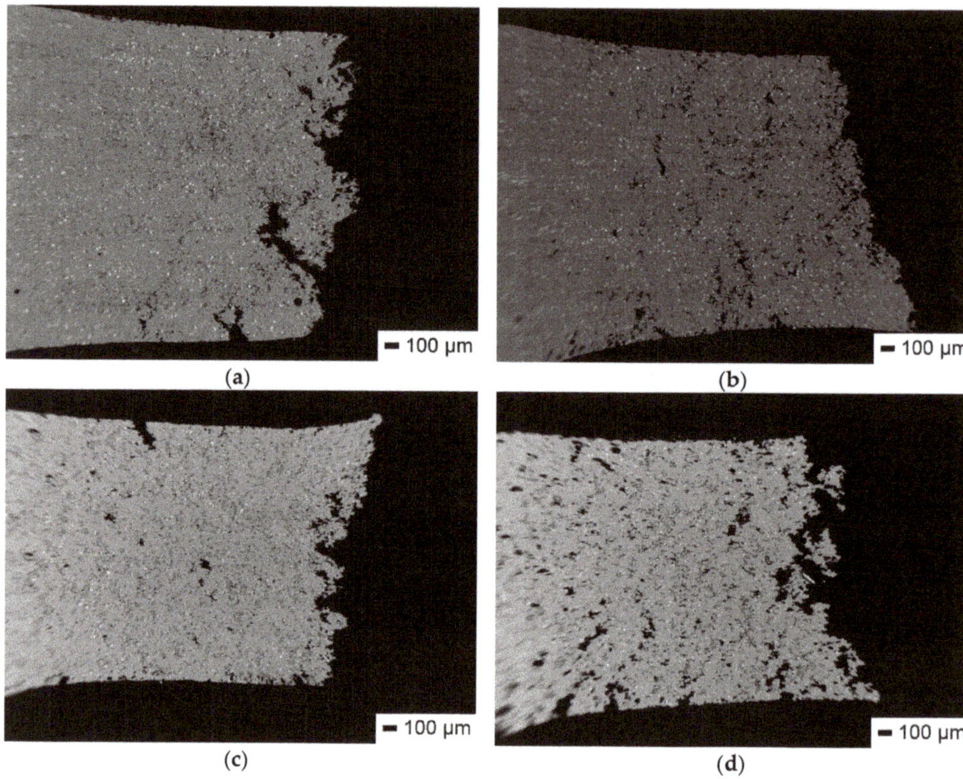

Figure 13. Fracture morphologies and cavity behaviors for the superplastic fraction surfaces of the 5A70 alloy at $T = 400$ °C, $\dot{\varepsilon} = 1 \times 10^{-3}$ s^{-1} (**a**); $T = 450$ °C, $\dot{\varepsilon} = 1 \times 10^{-3}$ s^{-1} (**b**); $T = 500$ °C, $\dot{\varepsilon} = 5 \times 10^{-4}$ s^{-1} (**c**); and $T = 550$ °C, $\dot{\varepsilon} = 5 \times 10^{-4}$ s^{-1} (**d**).

Figure 13a–d show that no significant necking occurred at the superplastic fracture, and cracks near the fracture location gradually spread outwards from the fracture along the accumulated deformation. Therefore, crack formation was the main reason for the ultimate fracture of superplastic tensile [75]. The pinning effect of the phase particles was strengthened to enhance the cavity nucleation and the chemical potential between the forceful grain boundary atoms and the free surface of the cavity. In a previous study, the results demonstrated a clear transition from diffusion growth to superplastic diffusion growth and plastic-controlled growth at a cavity radius larger than 1.52 and 13.90 μm [71]. Cavity growth mainly occurred at the stage of diffusion growth and superplastic diffusion growth due to the diffusion of the voids into adjacent cavities with a cavity radius smaller than the grain size. Plastic-controlled growth dominated the cavity interlinkage and coalescence process, which eventually led to superplastic fractures. Therefore, the irregular-shaped cavity with the accumulation of deformation strains induced wedge cracks, which propagated and converged under the shear stresses until a fracture occurred.

At the first stage, the crack in the specimen propagated as an opening mode crack that formed vortex structures (Figure 7). Second phase particles were formed along the cavity surface, which indicates a high mass transfer rate in the vortex porous structure. At the second stage, periodic transverse shear displacement occurred at the fracture surface (Figure 13b,c) due to a change of the crack type: the crack at this stage propagated as a sliding mode crack. As the crack reached the specimen edges, it rotated and propagated along the direction of τ_{max} under plane stresses. At this stage, the fracture fractograph revealed traces of material rotations along the longitudinal shear in the form of plane cavities and discontinuities (Figure 13d) [76].

At 550 °C, the dissolution of the phase particles reduced the suppression of grain growth during dynamic recrystallization. However, the grain growth promoted the healing of the small cavities [71]. This clearly verifies that the cavity nucleation and cavity growth during superplastic tensile deformation promoted steady-state flows. Ultimately, the cracks destroyed the superplastic tensile stability and resulted in superplastic fracturing with no obvious necking.

5. Conclusions

The following conclusions are drawn from this work:

(1) Static recrystallization at 340 °C for 20 min showed an average grain size of 8.48 μm for the 2 mm thick 5A70 alloy sheet, which was due to the pinning effect from the dispersion of the second phase particles.

(2) The 5A70 aluminum alloy exhibited reasonable δ values at 400 °C (205%) and 450 °C (321%) with a moderate strain rate (1×10^{-3} s^{-1}), while the corresponding coefficients for the strain rate sensitivity, m, were 0.42 and 0.40, respectively. However, excellent δ values were obtained with a low strain rate of 5×10^{-4} s^{-1} at 500 (398%) and 550 °C (437%) with m = 0.47 and 0.46, respectively.

(3) The strain hardening of the 5A70 alloy preferentially occurred at high temperatures (500 and 550 °C) with low strain rates (5×10^{-4} s^{-1}). This was due to the decreased Mg-rich phase particles, and the abnormal grain growth weakened the pinning effect during dynamic recrystallization.

(4) The activation energy at 400–550 °C ranged from 135 to 139 kJ/mol, which is close to the lattice diffusion activation energy of pure aluminum (143.4 kJ/mol). Therefore, lattice diffusion dominated the GBS mechanism of the 5A70 alloy during superplastic deformation.

(5) Massive second phase particles played a significant effect in suppressing the dynamic recrystallization during superplastic tensile and promoted the cavity nucleation, growth, interlinkage, and coalescence during superplastic deformation.

(6) The formation of submicrometer filaments at the fraction surfaces were due to the Mg-rich phase particles that precipitated along the grain boundaries in the GBS and from the growth of MgO and $MgAl_2O_4$ that formed at high temperatures during superplastic formation.

Author Contributions: Conceptualization, Z.H. and S.L.; methodology, S.L.; software, S.L.; validation, Z.H., S.J., and S.L.; formal analysis, S.L.; investigation, S.L.; resources, Z.H. and S.J.; data curation, S.L.; writing—original draft preparation, S.L.; writing—review and editing, S.L.; supervision, H.Z.; project administration, S.J.; funding acquisition, S.J.

Funding: This research was funded by the Fundamental Research Funds for the Central Universities of China, grant number FRF-TP-16-043Al and the APC was funded by S.J.

Acknowledgments: We would like to thank Guoliang Xie for help in material preparation, Xuetong Zhao for assistance in the superplastic tensile tests, Shuming Wang for assistance in the SEM/EBSD analysis, and Xinqin Liu for help analyzing the result.

Conflicts of Interest: The authors declare no conflict of interest.

References

1. Bolt, P.J.; Lamboo, N.A.P.M.; Rozier, P.J.C.M. Feasibility of warm drawing of aluminium products. *J. Mater. Process. Technol.* **2001**, *115*, 118–121. [CrossRef]
2. Naka, T.; Torikai, G.; Hino, R.; Yoshida, F. The effects of temperature and forming speed on the forming limit diagram for type 5083 aluminum–magnesium alloy sheet. *J. Mater. Process. Technol.* **2001**, *113*, 648–653. [CrossRef]

3. Lin, S.P.; Nie, Z.R.; Huang, H.; Li, B.L. Annealing behavior of a modified 5083 aluminum alloy. *Mater. Des.* **2010**, *31*, 1607–1612. [CrossRef]
4. Popović, M.; Romhanji, E. Stress corrosion cracking susceptibility of Al–Mg alloy sheet with high Mg content. *J. Mater. Process. Technol.* **2002**, *125–126*, 275–280. [CrossRef]
5. Tokuda, M.; Inaba, T.; Ohigashi, H.; Kurakake, A. Discussions on constitutive equations of superplastic 5083 aluminum alloy. *Int. J. Mech. Sci.* **2001**, *43*, 2035–2046. [CrossRef]
6. Langdon, T.G. Seventy-five years of superplasticity: Historic developments and new opportunities. *J. Mater. Sci.* **2009**, *44*, 5998–6010. [CrossRef]
7. Iwahashi, Y.; Horita, Z.; Nemoto, M.; Langdon, T.G. Factors influencing the equilibrium grain size in equal-channel angular pressing: Role of Mg additions to aluminum. *Metall. Mater. Trans. A* **1998**, *29*, 2503–2510. [CrossRef]
8. Portnoy, V.K.; Rylov, D.S.; Levchenko, V.S.; Mikhaylovskaya, A.V. The influence of chromium on the structure and superplasticity of Al–Mg–Mn alloys. *J. Alloys Compd.* **2013**, *581*, 313–317. [CrossRef]
9. Woo, S.S.; Kim, Y.R.; Shin, D.H.; Kim, W.J. Effects of Mg concentration on the quasi-superplasticity of coarse-grained Al-Mg alloys. *Scr. Mater.* **1997**, *37*, 1351–1358. [CrossRef]
10. Kishchik, A.A.; Mikhaylovskaya, A.V.; Levchenko, V.S.; Portnoy, V.K. Formation of microstructure and the superplasticity of Al–Mg-based alloys. *Phys. Met. Metallogr.* **2017**, *118*, 96–103. [CrossRef]
11. Chang, S.Y.; Lee, J.G.; Park, K.T.; Shin, D.H. Microstructures and Mechanical Properties of Equal Channel Angular Pressed 5083 Al Alloy. *Mater. Trans.* **2001**, *42*, 1074–1080. [CrossRef]
12. Patlan, V.; Higashi, K.; Kitagawa, K.; Vinogradov, A.; Kawazoe, M. Cyclic Response of Fine Grain 5056 Al-Mg Alloy Produced by Equal Channel Angular Pressing. *Mater. Sci. Eng. A* **2001**, *319–321*, 587–591. [CrossRef]
13. Patlan, V.; Vinogradov, A.; Higashi, K.; Kitagawa, K. Overview of fatigue properties of fine grain 5056 Al-Mg alloy processed by equal channel angular pressing. *Mater. Sci. Eng. A* **2001**, *300*, 171–182. [CrossRef]
14. Yamashita, A.; Yamaguchi, D.; Horita, Z.; Langdon, T.G. Influence of pressing temperature on microstructural development in equal-channel angular pressing. *Mater. Sci. Eng. A* **2000**, *287*, 100–106. [CrossRef]
15. Fakhar, N.; Saniee, F.F.; Mahmudi, R. Significant improvements in mechanical properties of AA5083 aluminum alloy using dual equal channel lateral extrusion. *Trans. Nonferrous Met. Soc. China* **2016**, *26*, 3081–3090. [CrossRef]
16. Park, K.T.; Hwang, D.Y.; Lee, Y.K.; Kim, Y.K.; Shi, D.H. High strain rate superplasticity of submicrometer grained 5083 Al alloy containing scandium fabricated by severe plastic deformation. *Mater. Sci. Eng. A* **2003**, *341*, 273–281. [CrossRef]
17. Charit, I.; Mishra, R.S. Evaluation of microstructure and superplasticity in friction stir processed 5083 Al alloy. *J. Mater. Res.* **2004**, *19*, 3329–3342. [CrossRef]
18. Charit, I.; Mishra, R.S. High strain rate superplasticity in a commercial 2024 Al alloy via friction stir processing. *Mater. Sci. Eng. A* **2003**, *359*, 290–296. [CrossRef]
19. Ma, Z.Y.; Mishra, R.S.; Mahoney, M.W. Superplastic deformation behaviour of friction stir processed 7075Al alloy. *Acta Mater.* **2002**, *50*, 4419–4430. [CrossRef]
20. Mahoney, M.W.; Barnes, A.J.; Bingel, W.H.; Fuller, C.B. Superplastic Forming of 7475 Al Sheet after Friction Stir Processing (FSP). *Mater. Sci. Forum* **2004**, *447–448*, 505–512. [CrossRef]
21. Salem, H.G.; Reynolds, A.P.; Lyons, J.S. Structural evolution and superplastic formability of friction stir welded AA 2095 sheets. *J. Mater. Eng. Perform.* **2004**, *13*, 24–31. [CrossRef]
22. Sato, E.; Itoh, G.; Takayama, Y.; Kitazono, K.; Morita, K.; Itoi, T.; Kobayashi, J. Superplastic Behavior of Friction Stir Welded Joints of an Al-Mg-Sc Alloy with Ultrafine Grained Microstructure. *Mater. Sci. Forum* **2016**, *838–839*, 338–343. [CrossRef]
23. Mironov, S.; Malopheyev, S.; Vysotskiy, I.; Kaibyshev, R. Superplasticity of Friction-Stir Welds of Zr-Modified 5083 Aluminum Alloy with Ultrafine-Grained Structure. *Defect Diffus. Forum* **2018**, *385*, 15–20. [CrossRef]
24. Beijing Institute of Aeronautical Materials Group. *Materials Technology of Aeronautics*; Aviation Industry Press: Beijing, China, 2014; pp. 131–143, ISBN 978-7-5165-0298-3.
25. Nieh, T.G.; Wadsworth, J.; Sherby, O.D. *Superplasticity in Metals and Ceramics*; Cambridge University Press: Cambridge, UK, 1997; pp. 22–30, ISBN 978-0-521-02034-3.
26. Yuzbekova, D.; Mogucheva, A.; Kaibyshev, R. Superplasticity of ultrafine-grained Al–Mg–Sc–Zr alloy. *Mater. Sci. Eng. A* **2016**, *675*, 228–242. [CrossRef]

27. Ota, S.; Akamatsu, H.; Neishi, K.; Furukawa, M.; Horita, Z.; Langdon, T.G. Low-Temperature Superplasticity in Aluminum Alloys Processed by Equal-Channel Angular Pressing. *Mater. Trans.* **2002**, *43*, 2364–2369. [CrossRef]
28. Xun, Y.W.; Mohamed, F.A. Superplastic behavior of Zn–22%Al containing nanoscale dispersion particles. *Acta Mater.* **2004**, *52*, 4401–4412. [CrossRef]
29. Prabu, S.B.; Padmanabhan, K.A. Superplasticity in and Superplastic Forming of Aluminum–Lithium Alloys. In *Aluminum-Lithium Alloy, Processing, Properties, and Applications*; Prasad, N.E., Gokhale, A.A., Wanhill, R.J., Eds.; Elsevier: Oxford, UK, 2014; Volume 8, pp. 221–258, ISBN 978-0-12-401698-9.
30. Lee, S.W.; Yeh, J.W. Superplasticity of 5083 alloys with Zr and Mn additions produced by reciprocating extrusion. *Mater. Sci. Eng. A* **2007**, *460–461*, 409–419. [CrossRef]
31. Langdon, T.G. A unified approach to grain boundary sliding in creep and superplasticity. *Acta Metall. Mater.* **1994**, *42*, 2437–2443. [CrossRef]
32. Stato, Y.S.; Park, S.H.C.; Michiuchi, M.; Kokawa, H. Constitutional liquation during dissimilar friction stir welding of Al and Mg alloys. *Scr. Mater.* **2004**, *50*, 1233–1236. [CrossRef]
33. Li, S.; Huang, Z.G.; Jin, S.Y.; Wang, B.Y.; Liu, X.Q.; Lei, K. Heat Treatment Process of Cold Rolled 5A70 Aluminum alloy Superplastic Sheet. *Chin. J. Rare Met.* **2018**, *42*, 283–288. [CrossRef]
34. Horita, Z.; Furukawa, M.; Nemoto, M.; Barnes, A.J.; Langdon, T.G. Superplastic forming at high strain rates after severe plastic deformation. *Acta Mater.* **2000**, *48*, 3633–3640. [CrossRef]
35. Smolej, A.; Klobčar, D.; Skaza, B.; Nagode, A.; Slaček, E.; Dragojević, V.; Smolej, S. Superplasticity of the rolled and friction stir processed Al–4.5 Mg–0.35Sc–0.15Zr alloy. *Mater. Sci. Eng. A* **2014**, *590*, 239–245. [CrossRef]
36. Lee, S.; Utsunomiya, A.; Akamatsu, H.; Neishi, K.; Furukawa, M.; Horita, Z.; Langdon, T.G. Influence of scandium and zirconium on grain stability and superplastic ductilities in ultrafine-grained Al–Mg alloys. *Acta Mater.* **2002**, *50*, 553–564. [CrossRef]
37. Mikhaylovskaya, A.V.; Yakovtseva, O.A.; Golovin, I.S.; Pozdniakov, A.V.; Portnoy, V.K. Superplastic deformation mechanisms in fine-grained Al–Mg based alloys. *Mater. Sci. Eng. A* **2015**, *627*, 31–41. [CrossRef]
38. Kulas, M.A.; Green, W.P.; Taleff, E.M.; Krajewski, P.E.; Mcnelley, T.R. Deformation mechanisms in superplastic AA5083 materials. *Metall. Mater. Trans. A* **2005**, *36*, 1249–1261. [CrossRef]
39. Yavari, P.; Langdon, T.G. An examination of the breakdown in creep by viscous glide in solid solution alloys at high stress levels. *Acta Mater.* **1982**, *30*, 2181–2196. [CrossRef]
40. McNalley, T.R.; Michel, D.J.; Salama, A. The Mg-concentration dependence of the strength of Al–Mg alloys during glide-controlled deformation. *Scr. Metall.* **1989**, *23*, 1657–1662. [CrossRef]
41. Gali, O.A.; Riahi, A.R.; Alpas, A.T. The effect of surface conditions on the elevated temperature sliding contact deformation of AA5083 alloy. *Wear* **2015**, *330–331*, 309–319. [CrossRef]
42. Cao, F.R.; Li, Z.L.; Zhang, N.X.; Ding, H.; Yu, F.X.; Zuo, L. Superplasticity, flow and fracture mechanism in an Al–12.7Si–0.7Mg alloy. *Mater. Sci. Eng. A* **2013**, *571*, 167–183. [CrossRef]
43. Shin, D.H.; Park, K.T. Directional cavity stringer formation in a superplastic 7075 Al alloy. *Mater. Sci. Eng. A* **1999**, *268*, 55–62. [CrossRef]
44. Blandin, J.J.; Hong, B.; Varloteaux, A.; Suery, M.; L'Esperance, G. Effect of the nature of grain boundary regions on cavitation of a superplastically deformed aluminium alloy. *Acta Mater.* **1996**, *44*, 2317–2326. [CrossRef]
45. Zelin, M.G. On micro-superplasticity. *Acta Mater.* **1997**, *45*, 3533–3542. [CrossRef]
46. Tan, J.C.; Tan, M.J. Superplasticity and grain boundary sliding characteristics in two stage deformation of Mg–3Al–1Zn alloy sheet. *Mater. Sci. Eng. A* **2003**, *399*, 81–89. [CrossRef]
47. Ritchie, I.M.; Sanders, J.V.; Weickhardt, P.L. Oxidation of a dilute aluminum magnesium alloy. *Oxid. Met.* **1971**, *3*, 91–101. [CrossRef]
48. Chang, J.K.; Taleff, E.M.; Krajewski, P.E.; Ciulik, J.R. Effects of atmosphere in filament formation on a superplastically deformed aluminum–magnesium alloy. *Scr. Mater.* **2009**, *60*, 459–462. [CrossRef]
49. Dirras, G. Filament formation during elevated temperature deformation of high purity ultrafine-grained aluminum. *Mater. Lett.* **2010**, *64*, 1163–1165. [CrossRef]
50. Das, S.; Riahi, A.R.; Meng-Burany, X.; Morales, A.T.; Alpas, A.T. High temperature deformation and fracture of tribo-layers on the surface of AA5083 sheet aluminum–magnesium alloy. *Mater. Sci. Eng. A* **2012**, *531*, 76–83. [CrossRef]

51. Taleff, E.M.; Nevland, P.J.; Krajewski, P.E. Tensile ductility of several commercial aluminum alloys at elevated temperatures. *Metall. Mater. Trans. A* **2001**, *32*, 1119–1130. [CrossRef]
52. Kaibyshev, R.; Sakai, T.; Nikulin, I.; Musin, F.; Goloborodko, A. Superplasticity in a 7055 aluminum alloy subjected to intense plastic deformation. *Mater. Sci. Technol.* **2003**, *19*, 1491–1497. [CrossRef]
53. Valiev, R.Z. Structure and mechanical properties of ultrafine-grained metals. *Mater. Sci. Eng. A* **1997**, *234–236*, 59–66. [CrossRef]
54. Mohamed, F.A. Interpretation of superplastic flow in terms of a threshold stress. *J. Mater. Sci.* **1983**, *8*, 582–592. [CrossRef]
55. Ma, Z.Y.; Liu, F.C.; Mishra, R.S. Superplastic deformation mechanism of an ultrafine-grained aluminum alloy produced by friction stir processing. *Acta Mater.* **2010**, *58*, 4693–4704. [CrossRef]
56. Watanabe, H.; Mukai, T.; Kohzu, M.; Tanabe, S.; Higashi, K. Effect of temperature and grain size on the dominant diffusion process for superplastic flow in an AZ61 magnesium alloy. *Acta Mater.* **1999**, *47*, 3753–3758. [CrossRef]
57. Sakai, T.; Miura, H.; Goloborodko, A.; Sitdikov, O. Continuous dynamic recrystallization during the transient severe deformation of aluminum alloy 7475. *Acta Mater.* **2009**, *57*, 153–162. [CrossRef]
58. Mao, W.; Yu, Y. Effect of elastic reaction stress on plastic behaviors of grains in polycrystalline aggregate during tensile deformation. *Mater. Sci. Eng. A* **2004**, *367*, 277–281. [CrossRef]
59. Sato, E.; Kuribayashi, K. Superplasticity and Deformation Induced Grain Growth. *ISIJ Int.* **1993**, *33*, 825–832. [CrossRef]
60. Cao, F.R.; Lei, F.; Cui, J.Z.; Wen, J.L. Modification of a deformation induced grain growth model of superplasticity and its experimental verification. *Acta Metall. Sin.* **1999**, *35*, 770–772.
61. Cao, F.R.; Ding, H.; Li, Y.L.; Zhou, G.; Cui, J.Z. Superplasticity, dynamic grain growth and deformation mechanism in ultra-light two-phase magnesium–lithium alloys. *Mater. Sci. Eng. A* **2010**, *527*, 2335–2341. [CrossRef]
62. Malopheyev, S.; Kipelova, A.; Nikulin, I.; Kaibyshev, R. Mechanical Properties of an Al-5.4%Mg-0.5%Mn-0.1%Zr Alloy Subjected to ECAP and Rolling. *Mater. Sci. Forum* **2011**, *667–669*, 815–820. [CrossRef]
63. Cao, F.R.; Cui, J.Z.; Wen, J.L.; Lei, F. Mechanical Behaviour and Microstructure Evolution of Superplastic Mg-8.4 wt pct Li Alloy and Effect of Grain Size and Phase Ratio on Its Elongation. *J. Mater. Sci. Technol.* **2000**, *16*, 55–58.
64. Cao, F.R. *Metal Superplasticity*; Metallurgical Industry Press: Beijing, China, 2014; pp. 211–212, ISBN 978-7-5024-6690-9.
65. Jie, J.C.; Zou, C.M.; Wang, H.W.; Li, B.; Wei, Z.J. Enhancement of mechanical properties of Al–Mg alloy with a high Mg content solidified under high pressures. *Scr. Mater.* **2011**, *64*, 588–591. [CrossRef]
66. Wang, G.S.; Liu, K.; Wang, S.L. Evolution of Elevated-Temperature Strength and Creep Resistance during Multi-Step Heat Treatments in Al-Mn-Mg Alloy. *Materials* **2018**, *11*, 1158. [CrossRef] [PubMed]
67. Zandbergen, H.W.; Andersen, S.J.; Jansen, J. Structure determination of Mg_5Si_6 particles in Al by dynamic electron diffraction studies. *Science* **1997**, *277*, 1221–1225. [CrossRef]
68. Wang, Y.; Liu, Z.K.; Chen, L.Q.; Wolverton, C. First-principles calculations of β''-Mg_5Si_6/α-Al interfaces. *Acta Mater.* **2007**, *55*, 5934–5947. [CrossRef]
69. Zhang, J.C.; Ding, D.Y.; Zhang, W.L.; Kang, S.H.; Xu, X.L.; Gao, Y.J.; Chen, G.Z.; Chen, W.G.; You, X.H. Effect of Zr addition on microstructure and properties of Al–Mn–Si–Zn-based alloy. *Trans. Nonferrous Met. Soc. China* **2014**, *24*, 3872–3878. [CrossRef]
70. Seifeddine, S.; Svensson, I.L. Prediction of mechanical properties of cast aluminium components at various iron contents. *Mater. Des.* **2010**, *31*, 6–12. [CrossRef]
71. Li, S.; Huang, Z.G.; Jin, S.Y. Cavity Behavior of Fine-Grained 5A70 Aluminum Alloy during Superplastic Formation. *Metals* **2018**, *8*, 1065. [CrossRef]
72. Bae, D.H.; Ghosh, A.K. Cavity growth during superplastic flow in an Al–Mg alloy: I. Experimental study. *Acta Mater.* **2002**, *50*, 993–1009. [CrossRef]
73. Takigawa, Y.; Aguirre, J.V.; Taleff, E.M.; Higashi, K. Cavitation during grain-boundary-sliding deformation in an AZ61 magnesium alloy. *Mater. Sci. Eng. A* **2008**, *497*, 139–146. [CrossRef]
74. Kawasaki, M.; Xu, C.; Langdon, T.G. An investigation of cavity growth in a superplastic aluminum alloy processed by ECAP. *Acta Mater.* **2005**, *53*, 5353–5364. [CrossRef]

75. Fan, G.J.; Choo, H.; Liaw, P.K.; Lavernia, E.J. Plastic deformation and fracture of ultrafine-grained Al–Mg alloys with a bimodal grain size distribution. *Acta Mater.* **2006**, *54*, 1759–1766. [CrossRef]
76. Panin, V.E.; Egorushkin, V.E.; Derevyagina, L.S.; Deryugin, E.E. Nonlinear wave processes of crack propagation in brittle and brittle-ductile fracture. *Phys. Mesomech.* **2013**, *16*, 183–190. [CrossRef]

metals

MDPI

Article

Effects of Prebending Radii on Microstructure and Fatigue Performance of Al-Zn-Mg-Cu Aluminum Alloy after Creep Age Forming

Dalian Yang [1,*], Jingjing Miao [1], Fanyu Zhang [1], Zhuo Fu [2] and Yilun Liu [3]

1 Hunan Provincial Key Laboratory of Health Maintenance for Mechanical Equipment, Hunan University of Science and Technology, Xiangtan 411201, China; mjjhkd@163.com (J.M.); zhangfanyu6@163.com (F.Z.)
2 Department of Mechanical and Electrical Engineering, Changsha University, Changsha 410022, China; fuzhuo2006@163.com
3 School of Mechanical and Electrical Engineering, Central South University, Changsha 410083, China; ylliu@csu.edu.cn
* Correspondence: ydlhnust@hnust.edu.cn; Tel.: +86-0731-58290624

Received: 2 April 2019; Accepted: 27 May 2019; Published: 30 May 2019

Abstract: A series of creep age forming experiments were carried out on 7075 aluminum alloy with different prebending radii at 443 K for 10 h. The hardness, tensile properties and high cycle fatigue performance of the alloy after creep age forming were obtained. In addition, the microstructure and fatigue fracture of the alloy were observed by transmission electron microscopy (TEM) and scanning electron microscope (SEM). The results show that with the increase of the prebending radius, the hardness, strength and elongation of the alloy increased; meanwhile, the conditional fatigue limit increased, the size of precipitated phase decreased and the quantity increased, while the width of precipitate free zone decreased.

Keywords: creep age forming; high cycle fatigue; microstructure; 7075 aluminum alloy

---

1. Introduction

The Al-Zn-Mg-Cu series alloys have been widely used in aerospace applications [1,2], such as for wing stringers, fuselage frames, wing skins, and other critical components owing to their high strength, low density, ability to be heat-treated, etc. With the rapid development of modern industry, the high requirements of aerospace equipment tend to accelerate the invention and the application of new manufacturing technologies. Creep age forming technology is one of those advanced aluminum alloy forming technologies, which synchronizes forming and artificial aging [3]. Compared with traditional forming methods such as drawing, rolling and bending, and shot peening, the creep age forming technology has a range of advantages such as lesser material waste than milling, lower residual stress and better surface quality. Additionally, the creep age forming technology can enhance the stress corrosion resistance of the alloy and extend the service life of parts [1,4]. Therefore, it is mainly used in the manufacture of large integral panel parts [5,6] and has a broad application prospect in the field of aircraft manufacturing [7–9].

In the past, scholars tended to study two main aspects of creep age forming technology: (1) the deformation control of aluminum alloy using the creep age forming technology (Peddieson [10] and Sallah [11] studied the relationship between creep and stress relaxation in the aging process based on viscoelastic mechanics. At the same time, they also gave the calculation formulas in their studies. Ho [12] established a unified creep/stress relaxation constitutive model that includes phase precipitation, grain growth and dislocations based on creep, stress relaxation theory and aging kinetics); and (2) improvement of microstructure and macroscopic performances of aluminum alloy using creep age forming technology. Sarioglu [13] and Brav [14] studied the fatigue crack propagation rate of 2024 aluminum alloy under different aging conditions, and found that aging treatment can improve the fatigue performance of the alloy. Lumley [15], Zhao [16] and Jin [7] reached similar conclusions. Temperature, time and stress are combined in the process of creep age forming, and are therefore regarded as the three important parameters of creep age forming technology. Yang [17] studied the effect of aging temperature on microstructure and fatigue performance. The results show that the fatigue performance of 7075 aluminum alloy increases with the increase of aging temperature; Zhan [18] tested the hardness, strength and elongation of 2124 aluminum alloy aged at 458 K under 0 MPa and 200 MPa, the results show that the strength of the alloy increases and the plasticity decreases under the stress aging condition. Zhu [19] studied the microstructure of Al-xCu aluminum alloy after aging and found that the yield strength of the alloy with stress was lower than that without stress; Liu [20] studied the strength and stress corrosion fracture (SCC) behavior of 7075 aluminum alloy after aging and regression treatment; the results showed that the strength, hardness and SCC sensitivity of 7075 aluminum alloy were closely related to aging temperature and time. Liu [21] studied the influence of aging temperature and time on the microstructure and mechanical properties of 7075 aluminum alloy sheet after creep age forming; the results showed that the creep age forming performance of 7075 aluminum alloy sheet is closely related to the process parameters.

In conclusion, during the investigation of forming microstructure and mechanical properties, the effects of temperature and time have been studied more than the effects of prebending radius on microstructure and high cycle fatigue (HCF) properties. As is well known, materials and structures exposed to HCF are subject to low cyclic stress, so the plastic deformation is not obvious and is difficult to detect and prevent. Therefore, the study of HCF performance is of great significance for improving the fatigue life of materials, in particular for 7075 aluminum alloy, a typical Al-Zn-Mg-Cu aluminum alloy, which has high strength and is widely applied in the aviation and aerospace industry. This paper outlines the findings of the study of the effects of prebending radii on the hardness, tensile strength, yield strength, elongation and HCF performance of the 7075 aluminum alloy after creep age forming. The microstructure and fatigue fracture morphology were further observed by scanning electron microscopy and transmission electron microscopy. The present research work can provide reference for the improvement of creep age forming technology and engineering application of 7075 aluminum alloy.

2. Materials and Methods

2.1. Material

The material used in experiment was the 7075 aluminum alloy plate with a thickness of 4 mm and a heat treatment temper of T651. The chemical compositions (in wt.%) were: 5.71% Zn, 2.45% Mg, 1.5% Cu, 0.18% Cr, 0.17% Fe, 0.06% Si, 0.034% Mn, 0.019% Ti, and a balance of Al.

2.2. Creep Age Forming Experiments

The 360 mm × 220 mm × 4 mm 7075 aluminum alloy plates were prepared by wire electrical discharge machining (WEDM). The solid solution treatment was carried out at 753 K for 0.5 h in a chamber type electric resistance furnace (SX-12-10, Beijing Ever Bright Medical Treatment Instrment Co., LTD., Beijing, China) and was immediately followed by quenching in water (298 K). The transfer time from the furnace to the water was less than 5 s.

The creep age forming experiment was conducted in an electro thermostatic blast oven (Shanghai Jing Hong Laboratory Instrument Co., LTD., Shanghai, China) which has a temperature accuracy of 0.1 K. The experiment rig is shown in Figure 1. The creep age forming experiment process includes three steps: (1) prebending—placing the 7075 aluminum alloy plate in the center of the mold, with a threaded rod and nut used to apply the prebending load, as shown in Figure 1a; (2) aging—the 7075 aluminum alloy plate was aged at 443 K for 10 h in the electro thermostatic blast oven, as shown in

Figure 1b; (3) unloading—removing the load and temperature after aging, the formed aluminum alloy plate was obtained after springback, as shown in Figure 1c. The experiment was divided into three groups, with the prebending radius ρ_1 = 500 mm, ρ_2 = 1000 mm, and ρ_3 = 1500 mm, respectively. In order to improve the credibility of the experiment, each group of experiments was carried out three times. The specific process parameters of the experiments are shown in Table 1.

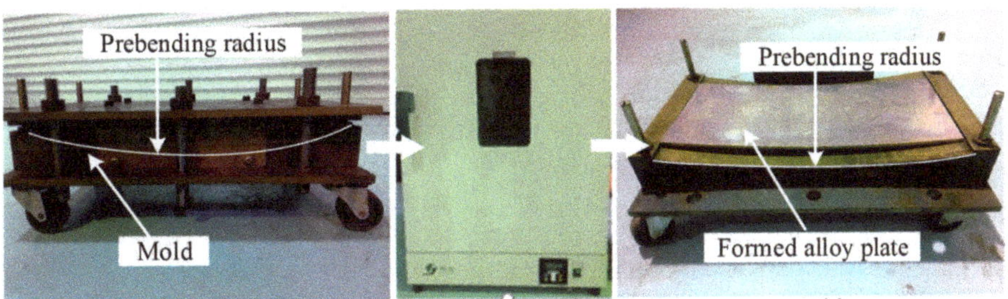

Figure 1. Creep age forming experiment: (**a**) prebending, (**b**) aging, and (**c**) unloading.

Table 1. The parameters of the experiments.

Group No.	Temperature/K	Time/h	ρ/mm
CT1	443	10	1500
CT2	443	10	1000
CT3	443	10	500

2.3. Specimen Preparation

After the creep age forming experiments, the tensile and HCF specimens of 7075 aluminum alloy were machined by WEDM along with the direction x (the rolling direction), y (the vertical rolling direction) and z (the thickness direction), as shown in Figure 2a. The shape and size of the specimens are shown in Figure 2b,c, respectively. Meanwhile, several specimens were machined for hardness testing and microstructure observations. All the specimen surfaces were polished using fine sand paper (1500 #) and rinsed with alcohol before testing.

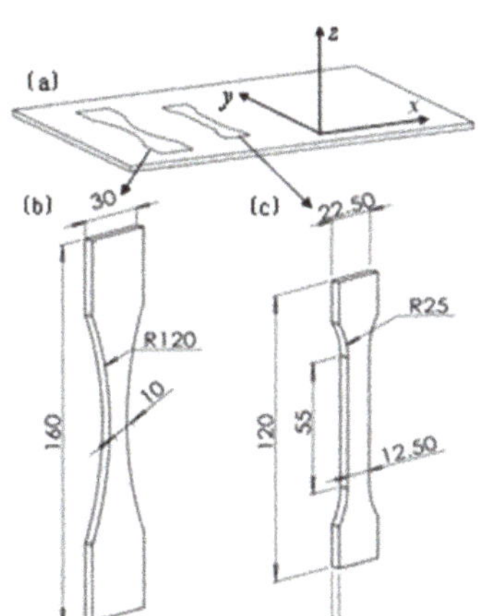

Figure 2. Size of the specimens: (**a**) cutting direction, (**b**) high cycle fatigue (HCF) specimen, and (**c**) tensile specimen (unit: mm).

2.4. Testing

The hardness was tested by a digital microhardness tester (HVS-1000Z, Shandong Laizhou Huayin Test Instrument Co., LTD., Shandong, China) at a load of 9.8 N with dwell time 30 s. Three different points were tested on each specimen, and the mean value was taken as the hardness of the specimen.

The mechanical properties were tested on a fully automated, closed-loop servo-hydraulic material testing system (MTS810-50 kN), including the ultimate tensile strength (UTS, σ_b), the 0.2% offset yield strength (YS, $\sigma_{0.2}$), and the elongation (δ). The mechanical properties were tested according with the specifications in ASTM standard E8M-1989. In this testing, three tensile specimens were tested for each group experiment, and the results were stated as the mean value. A high-sensitivity extensometer with a gauge distance of 20 mm was installed. The tensile elongation rate for all tests was 2 mm/min.

The S-N curves were tested on the material testing system (MTS810-50 kN) by the stress-control method. All HCF specimens were tested according with the specification of ASTM standard E466-2007. The stress ratio R was 0.1, the peak stress S_{max} was 160–320 MPa and the loading frequency was 40 Hz. The whole process was completed at room temperature (298 K) and air. The data were monitored and recorded automatically by the computer.

2.5. Microstructure Observation

The microstructure of the specimens was observed by transmission electron microscopy (Fei Tecnai G2 F20, United States FEI Limited Liability Company, Hillsboro, OR, USA) based on Schottky field emission. The acceleration voltage was 200 kV, and the magnification was 25,000–10,300,000×. The specimens were prepared according with the following steps: first, some discs with a thickness of 1 mm and a diameter of 3 mm were prepared; next, the discs were thinned to 50 μm by polishing; finally, the discs were grounded and thinned by electropolishing (MTP-1) using a 30% nitric acid in methanol solution as electrolyte. The temperature was kept between 153 K and 303 K by liquid nitrogen, and the voltage was set to 10–20 V, the current was set to 50–60 mA, and the polishing time was 2 min.

After the HCF tests, the fracture was completely cut off from the specimen and observed by a scanning electron microscope (TESCAN MORA3 LMU, Tescan Company, Brno, Czech Republic). The acceleration voltage was 0.2–30 kV and the magnification was 3.5–1,000,000×. The resolution can reach 1.0 nm in high vacuum mode.

3. Results and Discussion

3.1. Effects of Prebending Radii on Mechanical Properties

The mechanical properties of 7075 aluminum alloy aged at different radii of preloading are shown in Table 2, where it can be seen that with the increase of prebending radius ρ, the tensile strength σ_b, yield strength $\sigma_{0.2}$, and elongation δ of 7075 aluminum alloy increased. The results indicate that the large prebending radius can improve the mechanical properties of the alloy. These effects will be discussed in-depth in Section 3.3.

Table 2. Mechanical properties of 7075 aluminum alloy at different preloading radii.

Group No.	Prebending Radii ρ/mm	Average Hardness/HV	Average UTS σ_b/MPa	Average YS $\sigma_{0.2}$/MPa	Average δ/%
CT1	1500	187	569.6	505	12.6
CT2	1000	183	552.47	468.64	12.51
CT3	500	172	538.36	466.51	6.12
The standard deviation	-	7.77	15.64	21.63	3.72

3.2. Effects of Prebending Radii on HCF

As can be seen from the data in Figure 3, the S-N curves of 7075 aluminum alloy under different prebending radii have a significant influence on the fatigue life of the formed alloy. When the prebending radius ρ was 1500 mm, the fatigue life was the longest under the same stress level. Under the higher stress levels (>210 MPa), the fatigue life of the alloy formed with prebending radius 500 mm was longer than that with prebending radius 1000 mm, while under the lower stress levels (<210 MPa), the fatigue life of the alloy formed with the prebending radius 500 mm was the shortest. The results show that increasing the prebending radius contributes to reducing the damage of the alloy under low stress (<210 MPa) and to improving the HCF performance of 7075 aluminum alloy.

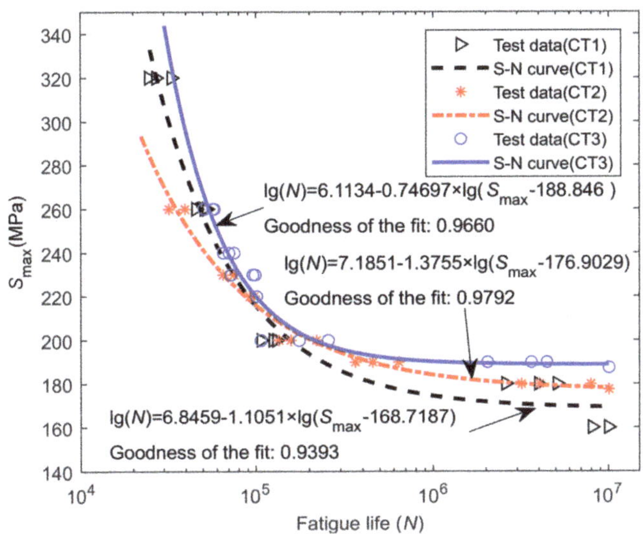

Figure 3. S-N curves of 7075 aluminum alloy at different prebending radii.

As aluminum and aluminum alloys are not commonly considered to have a fatigue limit, the conditioned fatigue limit is usually used in practical engineering applications. According to the method in literature [22], the conditioned fatigue limit is defined as the fatigue life when the number of cycles is 10^7. The conditional fatigue limits of 7075 aluminum alloy with different prebending radii are calculated and obtained from SN curves, as shown in Table 3, where the data also indicate that the prebending radius has a significant influence on the conditional fatigue limit of the alloy. The conditional fatigue limit of the alloy increased as the preload radius increased. When the prebending radius was 1500 mm, the conditional fatigue limit reaches the maximum value: 188.91 MPa.

Table 3. The conditional fatigue limits of 7075 aluminum alloy at different prebending radii.

Group No.	Prebending Radii/mm	Conditional Fatigue Limits/MPa
CT1	1500	188.91
CT2	1000	178.27
CT3	500	169.44

3.3. Effects of Prebending Radii on Microstructure

The TEM images of 7075 aluminum alloy at different prebending radii are shown in Figure 4, where the data indicate that the prebending radii have a significant influence on the microstructure of 7075 aluminum alloy. By measuring over 100 precipitates in TEM images, the average precipitate size was 60 nm in width and 80 nm in length in the matrix, which was the largest at a prebending radius of 500 mm, and the width of the precipitate free zone (PFZ) was the widest, approximately 100 nm, as shown in Figure 4c. The PFZ width and precipitate size decrease as the prebending radius increases. When the prebending radius was 1000 mm, the width of PFZ was 81 nm, and the average precipitate size was 55 nm in width and 64 nm in length, as shown in Figure 4b. When the prebending radius was 1500 mm, the width of PFZ was only half of that when the radius was 500 mm, and the average precipitate size was 50 nm in width and 60 nm in length, as shown in Figure 4a.

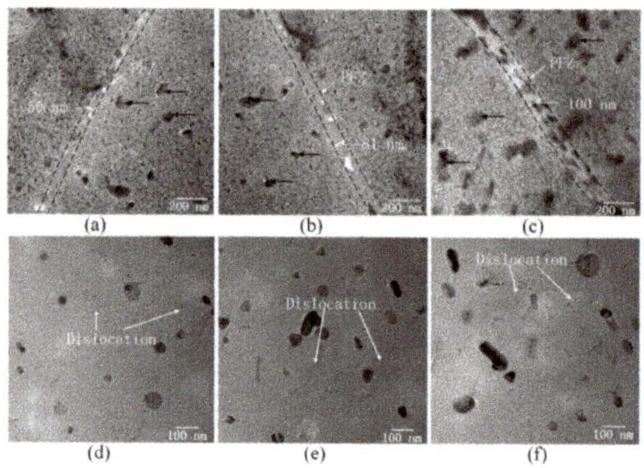

Figure 4. TEM microstructure of 7075 aluminum alloy at different prebending radii. (**a**,**d**) 1500 mm; (**b**,**e**) 1000 mm; (**c**,**f**) 500 mm.

Previous studies [23] have shown that stress can increase dislocation density, thus accelerating the rate of phase transformation and growth. In addition, the dislocations have a significant effect on fatigue performance: the higher the number of dislocations, the greater the interface energy, and the greater the stress concentration around the dislocations. As can be seen from Figure 4, at a small stress (the prebending radius was 1500 mm), the precipitated phase in the matrix is a finer dispersion than in the other cases. On the one hand, the smaller the size and the more dispersed the precipitated phase, the stronger the combining capacity between the precipitated phase and the matrix, which can result in stronger fracture toughness, hardness and strength [24]. In this experiment, with the increase of the prebending radius, the size of the precipitated phase becomes smaller and the distribution becomes more homogeneous. Therefore, the hardness and tensile properties of the alloy increased. On the other hand, the smaller the size and the more dispersed the precipitated phase, the better the fatigue performance.

In this experiment, the average number and average length of dislocations were measured on an area of 0.93 μm × 0.93 μm of TEM images more than ten times. The stress was the largest at prebending radius of 500 mm, and thus the average density of dislocations was 7.996×10^9 cm^{-2} and their average length (77.1 nm) of dislocations were also the largest, as shown in Figure 4f. This, together with the fact that the size of the precipitated phase was also the largest, causes the conditional fatigue limit to be the lowest. The average density of dislocations was 5.239×10^9 cm^{-2} and the average length was 58.6 nm when the prebending radius was 1000 mm, as shown in Figure 4e. When the prebending radius was 1500 mm, the average density of dislocations was 3.86×10^9 cm^{-2} and the average length (49 nm) of dislocations were smaller than those at the prebending radius of 500 mm and 1000 mm, and the size of precipitated phase was smallest as well, as shown in Figure 4d, so the fatigue life was the highest.

3.4. Effects of Prebending Radii on Fracture Morphology

Under 260 MPa, the fatigue fracture topographies of 7075 aluminum alloy after creep age forming with different prebending radii are shown in Figure 5. The data show that there were three typical regions on fatigue fracture topographies, which are marked by A, B and C, where A is the fatigue crack initiation region, B is the fatigue stable propagation region, and C is the fatigue transient fracture region. Some shear lips appeared at the edge of the specimens, which are represented by the white arrow mark.

Under 260 MPa, the morphologies of the fatigue crack initiation regions of 7075 aluminum alloy after creep age forming with different prebending radii are shown in Figure 6. The data show that radial stripes appear in the fatigue crack initiation regions under low magnification (Figure 6a–c), while under high magnification (Figure 6d–f), many fine steps appear in the fatigue initiation regions, which have smooth surface and many tiny cracks. Those steps are the secondary cracks formed in the nucleation process of the crack, but which lose power.

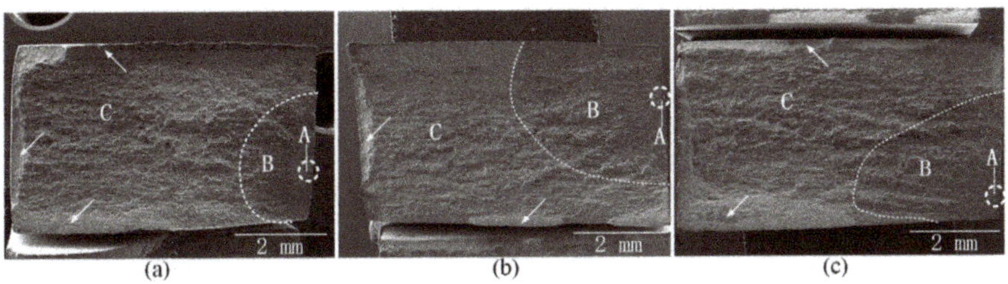

Figure 5. Fatigue fracture topography of 7075 aluminum alloy at different prebending radii (S_{max} = 260 MPa). (**a**) 1500 mm; (**b**) 1000 mm; (**c**) 500 mm.

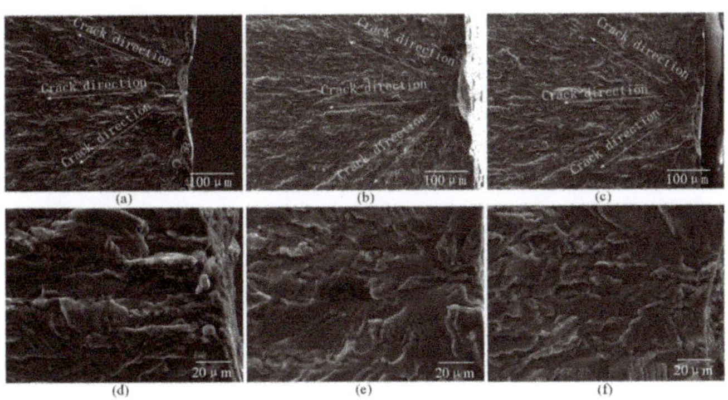

Figure 6. SEM morphologies of fatigue crack initiation regions of the 7075 aluminum alloy at different prebending radii (S_{max} = 260 MPa). (**a**,**d**) 1500 mm; (**b**,**e**) 1000 mm; (**c**,**f**) 500 mm.

Under 260 MPa, the SEM morphologies of the fatigue stable propagation regions of 7075 aluminum alloy after creep age forming with different prebending radii are shown in Figure 7. The data indicate that the fracture shows many fatigue striations and furrows. When the prebending radius was 1500 mm, the average width of the fatigue striation was 0.426 μm, obtained by measuring over 5 strips in SEM images. The average fatigue striation width was 0.308 μm when the prebending radius was 1000 mm and 0.368 μm when the prebending radius was 500 mm.

Under 260 MPa, the SEM images of the fatigue transient fracture regions of 7075 aluminum alloy after creep age forming with different prebending radii are shown in Figure 8. As can be seen from Figure 8, the fracture transient fracture regions show that there were both intergranular fractures and transgranular fractures. Intergranular fracture forms the stratified fracture planes on the fracture surface, while transgranular fracture forms dimples on the fracture surface [25]. When the prebending radius was 1500 mm, the dimples of 7075 aluminum alloy fracture were the largest. As the prebending radius decreases, the dimples become progressively smaller. It indicates that the toughness of the alloy decreases with the decrease of the prebending radius, which was consistent with the decrease in the elongation of the alloy.

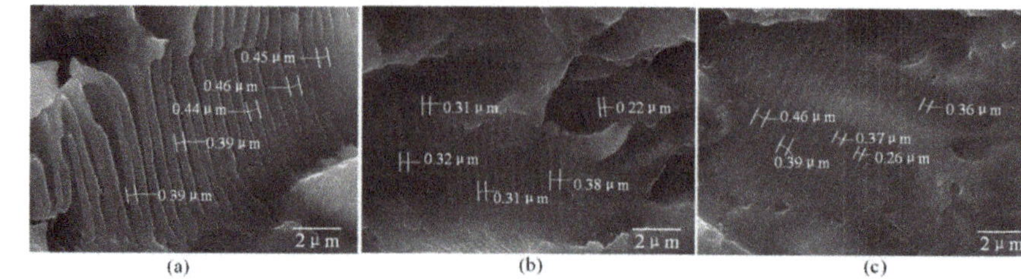

Figure 7. SEM morphologies of fatigue stability propagation regions of 7075 aluminum alloy at different prebending radii (S_{max} = 260 MPa). (**a**) 1500 mm; (**b**) 1000 mm; (**c**) 500 mm.

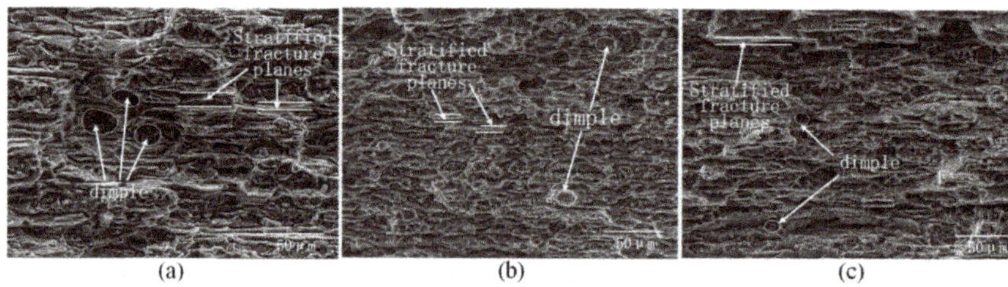

Figure 8. SEM morphologies of fatigue transient fracture region of 7075 aluminum alloy under different prebending radii (S_{max} = 260 MPa). (**a**) 1500 mm; (**b**) 1000 mm; (**c**) 500 mm.

4. Conclusions

In this paper, three groups of creep age forming experiments of 7075 aluminum alloy under different prebending radii were carried out, and the effects of three prebending radii (500 mm, 1000 mm, and 1500 mm) on the mechanical properties, HCF performance and microstructure of 7075 aluminum alloy after forming were studied. The following conclusions were drawn:

(1) Under the condition of 443 K + 10 h, with the increase of the prebending radius, the hardness, strength, and elongation increased. When the prebending radius was 1500 mm, the maximum value was 187 HV, σ_b = 568.98 MPa, $\sigma_{0.2}$ = 466.51 MPa, δ = 6.12%.

(2) The conditional fatigue limit increases with the increase of the prebending radius. When the prebending radius was 1500 mm, the maximum value was 188.91 MPa.

(3) With the increase of prebending radius, the size and density of precipitated phase become smaller and larger, and the width of PFZ also becomes narrower. In this experiment, when the preloading radius was 500 mm, the width of the PFZ was the largest (100 nm), while when the preloading radius was 1500 mm, the width of the PFZ was the smallest (50 nm).

Author Contributions: D.Y. and Y.L. conceived and designed the experiment; D.Y., J.M. and F.Z. performed the experiments; D.Y., J.M., and F.Z. analyzed the data; Z.F. and Y.L. contributed reagents, materials, and analysis tools; D.Y. wrote and edited the paper; all authors were involved in discussion for the paper.

Funding: This research was funded by the Project Supported by Scientific Research Fund of Hunan Provincial Education Department, grant number 17C0638; National Natural Science Foundation of China, grant number 11702091 and the Natural Science Foundation of Hunan Province of China, grant number 2018JJ3140; Science and Technology Foundation of Changsha, grant number k1705052.

Conflicts of Interest: The authors declare no conflict of interest.

References

1. Zhu, A.W.; Starke, E.A. Materials aspects of age-forming of Al-*x*Cu Alloys. *J. Mater. Process. Technol.* **2001**, *117*, 354–358. [CrossRef]
2. Inforzato, D.J.; Costa, J.P.R.; Fernandez, F.F.; Travessa, D.N. Creep-age forming of AA7475 aluminum panels for aircraft lower wing skin application. *Mater. Res.* **2012**, *15*, 596–602. [CrossRef]

3. Liu, C.; Liu, Y.; Li, S.B.; Ma, L.Y.; Zhao, X.Q.; Wang, Q. Effect of creep aging forming on the fatigue crack growth of an AA2524 alloy. *Mater. Sci. Eng. A* **2018**, *725*, 375–381. [CrossRef]
4. Brewer, H. Age forming integrally stiffened, aluminum aerospace structures in an autoclave. In Proceedings of the Aircraft Design & Operations Meeting, Seattle, WA, USA, 31 July–2 August 1989.
5. Pitch, E.P.D.; Styles, C.M. Creep age forming of 2024, 8090 and 7449 alloys. *Mater. Sci. Forum* **2000**, *331–337*, 455–460. [CrossRef]
6. Huang, X.; Zeng, Y.S. Study on spring back during creep age forming of aluminum alloy 7075. *J. Plast. Eng.* **2012**, *19*, 79–82. [CrossRef]
7. Jin, X.; Wan, M.; Li, C. Effect of creep age forming on fatigue for 7B04 aluminum alloy. *Forg. Stamp. Technol.* **2011**, *26*, 124–127. [CrossRef]
8. Adachi, T.; Kimura, S.; Nagayama, T. Age forming technology for aircraft wing skin. *Mater. Forum* **2004**, *28*, 202–207.
9. Eberl, F.; Gardineer, S.; Campanile, G. Age formable panels for commercial aircraft. *Proc. Inst. Mech. Eng. Part G Aerosp. Eng.* **2008**, *222*, 872–886. [CrossRef]
10. Peddieson, J.; Buchanan, G.R. Mathematical modeling of an age-forming process. *Math. Comput. Model.* **1990**, *14*, 1057–1060. [CrossRef]
11. Sallah, M.H.; Peddieson, J.; Foroudastan, S. A mathematical model of autoclave age forming. *J. Mater. Process. Technol.* **1991**, *28*, 211–219. [CrossRef]
12. Ho, K.C.; Lin, J.; Dean, T.A. Constitutive modeling of primary creep for age forming an aluminum alloy. *J. Mater. Process. Technol.* **2004**, *153*, 122–127. [CrossRef]
13. Sarioglu, F.; Orhaner, F.O. Effect of prolonged heating at 130 °C on fatigue crack propagation of 2024 Al alloy in three orientations. *Mater. Sci. Eng. A* **1998**, *248*, 115–119. [CrossRef]
14. Bray, G.H.; Glazov, M.; Rioja, R.J.; Li, D.; Gangloff, R.P. Effect of artificial aging on the fatigue crack propagation resistance of 2000 series aluminum alloys. *Int. J. Fatigue* **2001**, *23*, S265–S276. [CrossRef]
15. Lumley, R.N.; O'Donnell, R.G.; Polmear, I.J. Enhanced fatigue resistance by under aging an Al-Cu-Mg-Ag alloy. *Mater. Forum* **2005**, *29*, 256–261.
16. Chen, G.-Q.; Fu, X.-S.; Zhao, F.; Zhou, W.-L. Microstructures and mechanical properties of 2A12 aluminum alloy after age forming. *Trans. Nonferrous Met. Soc. China* **2012**, *22*, 1975–1980. [CrossRef]
17. Yang, D.L.; Liu, Y.L.; Li, S.B.; Ma, L.Y.; Liu, C.; Yi, J.H. Effects of aging temperature on microstructure and high cycle fatigue performance of 7075 aluminum alloy. *J. Wuhan Univ. Technol. (Mater. Sci.)* **2017**, *32*, 677–684. [CrossRef]
18. Zhan, L.H.; Li, Y.G.; Huang, M.H. Microstructures and properties of 2124 alloy creep ageing under stress. *J. Cent. South Univ. (Sci. Technol.)* **2012**, *43*, 926–931.
19. Zhu, A.W.; Starke, E.A. Stress ageing of Al-xCu alloys: Experiments. *Acta Mater.* **2001**, *49*, 2285–2295. [CrossRef]
20. Liu, J.H.; Li, D.; Liu, P.Y.; Guo, B.L.; Zhu, G.W. Effect of ageing and retrogression treatments on mechanical and corrosion properties of 7075 aluminum alloy. *Trans. Met. Heat Treat.* **2002**, *23*, 50–53. [CrossRef]
21. Liu, D.H.; Xie, Y.X.; Li, J.C. Influence of process parameters on creep age forming formability of 7075 aluminum alloy sheets. *Mater. Sci. Technol.* **2015**, *23*, 50–56. [CrossRef]
22. Yang, D.L.; Liu, Y.L.; Li, S.B.; Yi, J.H.; Tao, J. Investigation of uniaxial asymmetric high-cycle fatigue failure behavior of 7075-T651 Aluminum Alloy Sheet. In Proceedings of the International Conference on Material Science and Applications (ICMSA 2015), Suzhou, China, 13–14 June 2015; Atlantis Press: Paris, France. [CrossRef]
23. Li, C.; Dai, S.L.; Zhang, K.; Ru, J.G. Effect of stress on microstructure and mechanical properties during age forming process of 7050 aluminum alloy. *Hangkong Cailiao Xuebao J. Aeronaut. Mater.* **2013**, *33*, 19–23. [CrossRef]
24. Wang, C.H.; Zhang, T.; Yin, H.X.; Lv, Z.F. Effect of aging treatment on microstructure and properties of 7075 aluminum alloy. *Heat Treat. Met.* **2017**, *42*, 87–90. [CrossRef]
25. Suresh, S. *Fatigue of Materials*; Cambridge University Press: Cambridge, UK, 1998; pp. 238–248. [CrossRef]

metals

MDPI

Article

Surface Quality Improvement of AA6060 Aluminum Extruded Components through Liquid Nitrogen Mold Cooling

Andrea Francesco Ciuffini [1,*], Silvia Barella [1], Cosmo Di Cecca [1], Andrea Gruttadauria [1], Carlo Mapelli [1], Luca Merello [1], Giacomo Mainetti [2] and Massimo Bertoletti [2]

1 Dipartimento di Meccanica, Politecnico di Milano, via La Masa 34, 20156 Milano, Italy, silvia.barella@polimi.it (S.B.); cosmo.dicecca@polimi.it (C.D.C.); andrea.gruttadauria@polimi.it (A.G.); carlo.mapelli@polimi.it (C.M.); mezzello@hotmail.it (L.M.)

2 A.t.i.e. Uno Informatica S.r.l., via Macon 30, 23900 Lecco, Italy, giacomo.mainetti@unoi.it (G.M.); massimo.bertoletti@unoi.it (M.B.)

* Correspondence: andreafrancesco.ciuffini@polimi.it; Tel.: +39-340-1253162

Received: 5 May 2018; Accepted: 29 May 2018; Published: 1 June 2018

Abstract: 6xxx aluminum alloys are suitable for the realization of both structural applications and architectural decorative elements, thanks to the combination of high corrosion resistance and good surface finish. In areas where the aesthetic aspects are fundamental, further improvements in surface quality are significant. The cooling of the extrusion mold via internal liquid nitrogen fluxes is emerging as an important innovation in aluminum extrusion. Nowadays, this innovation is providing a large-scale solution to obtain high quality surface finishes in extruded aluminum semi-finished products. These results are also coupled to a significant increase in productivity. The aim of the work is to compare the surface quality of both cooled liquid nitrogen molds and classically extruded products. In this work, adhesion phenomena, occurring during the extrusion between the mold and the flowing material, have been detected as the main causes of the presence of surface defects. The analysis also highlighted a strong increase in the surface quality whenever the extrusion mold was cooled with liquid nitrogen fluxes. This improvement has further been confirmed by an analysis performed on the finished products, after painting and chromium plating. This work on the AA6060 alloy has moreover proceeded to roughness measurements and metallographic analyses, to investigate the eventual occurrence of other possible benefits stemming from this new extrusion mold cooling technology.

Keywords: aluminum; extrusion; liquid nitrogen cooling; surface quality; aesthetic

1. Introduction

Extrusion is one of the most widely used forming processes for aluminum and its alloys [1,2]. In this process, the internal liquid nitrogen cooling of the extrusion mold represents a significant innovation of recent years [3,4]. Indeed, during the extrusion process, different thermal exchanges occur, for example, heat generation due to material deformation, heat generation due to the friction between the billet and the liner and between the material and the dye, heat generation due to the internal friction between the flowing material and the dead-metal zone, and heat conduction through the components of the system (billet, ram, chamber, dye, etc.). Furthermore, due to the high output speed of this forming technology, the output speed of the is semi-product also high. Thus, the contact time between the mold and the extruded material is very short and the temperature increments are concentrated within a thin surface layer. This will extensively increase the generation of surface defects [5–10].

The new liquid nitrogen cooling technology, used directly in the matrix, allows the exploitation not only of the nitrogen inertizing effect, but also of its use as coolant, providing a much broader range of benefits. The liquid nitrogen enters into the channels inside the matrix at a temperature of −196 °C and, during evaporation, it cools the mold. After the change in state, the gas creates an inert atmosphere, which inhibits the formation of oxides. The most significant advantage of this process, which justifies the higher installation and operating costs, is the increase in productivity due to the decrease in heat generation during extrusion. Indeed, the modern liquid nitrogen extrusion mold cooling systems optimize the consumption of nitrogen and guarantee sufficiently precise control over the whole process [3,4,11–13]. A schematic sketch of this technology is reported in Figure 1.

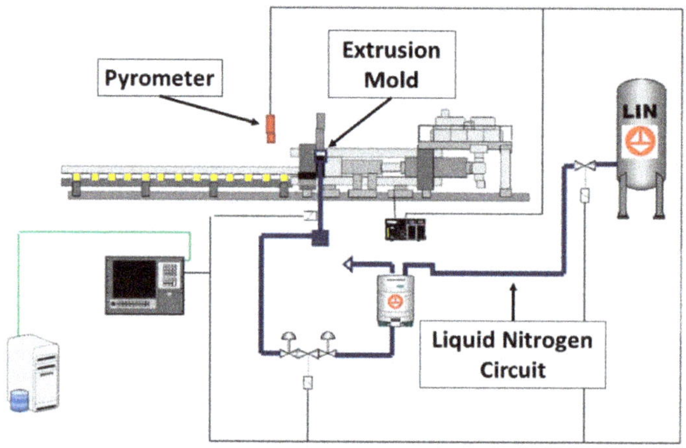

Figure 1. Schematic drawing of the liquid nitrogen extrusion mold cooling.

Furthermore, both the corrosion resistance and the aesthetic aspect of the final products may be relevant in some application fields of aluminum alloy extruded semi-products. Thus, surface quality may prove to be crucial [14–17].

Since the temperature increments of the extrusion process are concentrated in a thin surface layer, they drastically affect the surface finishing [2,4,18,19]. It is thus clear that the use of liquid nitrogen cooling of the mold and good temperature control of the overall process will beneficially affect these properties as well.

In the present work, the AA6060 alloy (0.30–0.60 wt % Si, 0.35–0.60 wt % Mg and 0.1–0.3 wt % Fe) is tested. This aluminum alloy is extremely ductile and therefore, is one of the most commonly used aluminum alloys. Further, its mechanical properties can be adjusted by subsequent heat treatment through precipitation hardening. It is used in many technical applications, such as automotive, aerospace and structural frame structures [20–22].

This work aims to assess the occurrence of a relationship between the surface quality and the use of internal liquid nitrogen cooling of the extrusion mold. Indeed, for structural frame applications, such as in this case, the extruded surface can be threated via the following aesthetic surface processes that might highlight surface defects. In the present work, in detail, the extruded semi-products undergo a 20 µm painting process and a 1 µm thick chromium plating process (Figure 2).

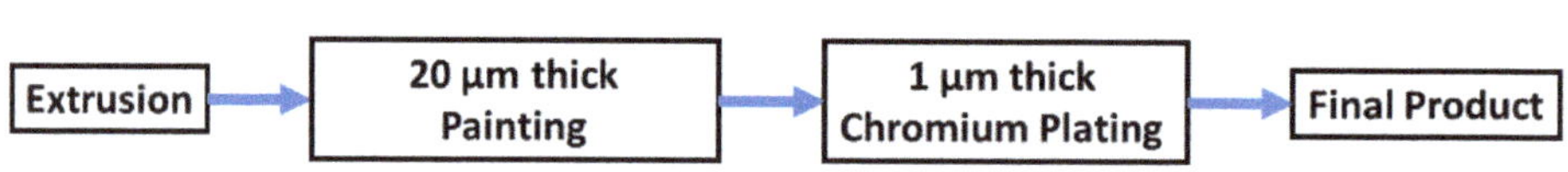

Figure 2. Flowchart of the production process of extruded AA6060 structural frames.

Thus, all the defects (such as pick-up, dye-lines and blisters), which are not completely hidden within these two overlaying layers, are highlighted in the final products by these aesthetic processes.

The surface finishing can accordingly prove a major instance in this application field and even small progresses could emerge as significant.

2. Experimental Procedure

During this work, two main sets of data were collected—one during the production trials of forty semi-finished extruded products, and the other one through laboratory analysis, performed to assess their quality.

The process parameters were monitored by process-control software, which manages data about material flow speed, temperature and the liquid nitrogen valve opening. In detail, data about dye temperature was collected through four thermocouples (TC) installed within the mold (Figure 3) and two laser pyrometers.

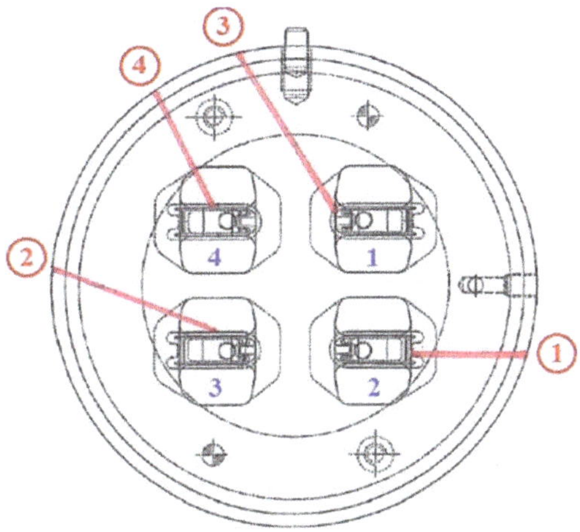

Figure 3. Schematic drawing of the thermocouples' positions within the dye.

The thermocouples used in this work were Chromel/Alumel (K) thermocouples with mineral oxide insulation and Inconel 600 as the sheath material. The sheath had a 3.2 mm diameter and 2000 mm length. The laser pyrometer temperature method measured the emitted thermal radiation, which directly corresponded to the temperature and surface emissivity of the target. In detail, the pyrometer sensors detected the amount of infrared radiation emitted by the measured object. The pyrometers were installed in front of the exit of the extrusion dye (Figure 4).

Figure 4. Pyrometers' positions.

On the other hand, the surface defects of the extruded semi-products were characterized in the junction zone of the billets, since this is an accumulation area for the defects [1,2]. The samples were drawn from the extruded semi-products and polished following the common standards. The chemical compositions of the samples were acquired through a scanning electron microscope (SEM) (Carl Zeiss AG, Oberkochen, Germany) Zeiss EVO 50®, using a back-scattered electron (SEM-BSE) (INCA Oxford Instrument, Oxford, UK) detector and energy dispersive X-ray microanalysis (SEM-EDS) (INCA Oxford Instrument, Oxford, UK). In addition, the same instrument was exploited to collect morphological images through a secondary electron detector (SEM-SE) (INCA Oxford Instrument, Oxford, UK).

The samples were etched electrochemically using a Barker's reagent. The electrolytic solution was prepared by mixing 40 parts of water (H_2O) for each volume-part of tetrafluoboric acid (HBF_4). The electrochemical anodization was performed by applying a current density of 0.2 A/cm^2 (20 V dc) for 40–80 s.

The metallographic analysis was executed using a cross polarized incident light. Afterwards, the index used to evaluate the average grain dimensions was calculated via the Heyn intercept method, following the ASTM E112-13 standard.

Roughness measurements were performed on the surface of the body of the extruded semi-product, hence, not in the junction area. Profile roughness measurements were taken perpendicularly to the extrusion direction, using a stylus Mahr PGK MFK-250® (Mahr GmbH, Göttingen, Germany) tester with a tip radius of 2 µm and a vertical measure range of +/−250 µm. The data were collected following the standard, UNI EN ISO 4288-2000. The translational speed of the measurement was 0.5 mm/s along an exploration length of 5.6 mm. The base length was 0.8 mm and the total evaluated length was 4.0 mm. The collected data were filtered using a Gaussian filter with a wavelength cutter of 0.8 mm.

Surface texture measurements were performed using an Alicona InfiniteFocus® (Alicona Imaging GmbH, Graz, Austria) microscope, which exploits focus-variation technology. The data were acquired in a similar position to those collected for the profile roughness measurements. The surface texture measurements parameters are listed as follows: the total evaluated area measured 4.0 × 4.0 mm and the base length was 0.8 mm. The collected data were filtered again using a Gaussian filter with a wavelength cutter of 0.8 mm.

3. Results

The data collected during the extrusion trials are summarized in Table 1. This sets out the data regarding the most meaningful specimens that were analyzed in this study. It should be underlined that the samples numbered from 1 to 15 were processed through a conventional extrusion, and then, the liquid nitrogen valve was opened and the specimens numbered up to 25 were subjected to transient conditions. Thus, only the semi-products numbered from 26 to 40 were extruded with the liquid nitrogen mold cooling system working at full capacity.

Table 1. Data featuring the extrusion trials of the specimens. Samples 5, 9, 13 were extruded conventionally. Samples 21 and 25 were processed during the transition step. Specimens 33 and 37 were extruded when the liquid nitrogen mold cooling worked at full capacity. The thermocouples (TC) were positioned within the mold as reported in Figure 3.

Sample	Billet Average Speed (m/min)	Billet Average Temperature (°C)	TC1 (°C)	TC2 (°C)	TC3 (°C)	TC4 (°C)	Liquid Nitrogen Valve Aperture (%)
5	15.94	526	518	507	497	527	0
9	17.06	534	527	514	503	534	0
13	19.62	545	528	516	505	538	0
21	22.18	528	504	482	482	522	60.46
25	16.74	528	479	456	450	505	20.05
33	15.64	511	474	447	447	497	11.72
37	15.51	506	463	439	441	485	27.08

The data regarding the characterization of defects are reported below. Two different species of defects were detected on the surface in the junction area of the extruded semi-products. Their SEM-BSE images are shown in Figure 5, and the acquired chemical compositions are quoted in Table 2.

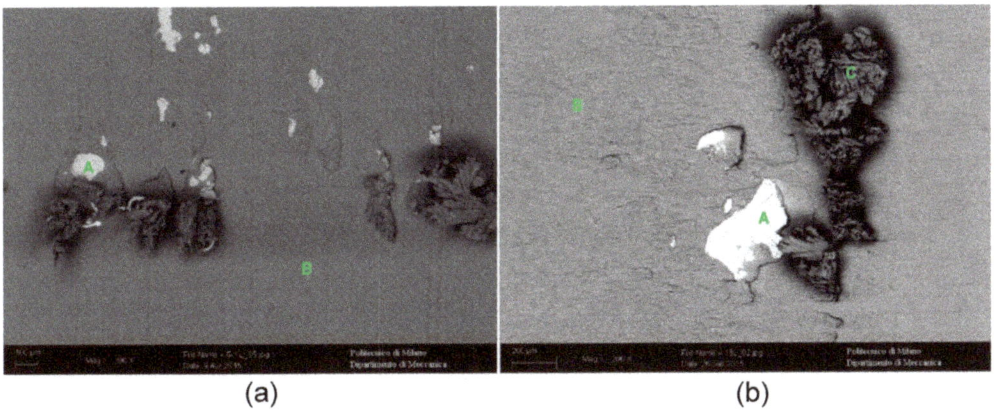

Figure 5. SEM-BSE images of extrusion defects detected in the junction area of the billets: sample 5 (**a**), sample 13 (**b**). A, B, C are the point in which the chemical analysis was performed (Table 2).

Table 2. SEM-EDS chemical analysis of the areas highlighted in Figure 5, the element concentrations are expressed as wt %.

Area	% Mg	% Al	% Si	% S	% V	% Cr	% Mo	% Fe
5-1 L_05_A	-	3.89	2.50	1.47	2.11	9.67	-	80.36
5-1 L_05_B	0.58	98.87	0.55	-	-	-	-	-
13-3 L_02_A	-	2.38	2.67	-	1.73	11.26	3.12	78.76
13-3 L_02_B	0.57	97.64	0.49	-	-	-	-	-
13-3 L_02_C	0.75	98.68	0.57	-	-	-	-	-

Therefore, the analyzed defects can be classified as pick-up and dye pick-up. Pick-ups (Figure 5a—point C) are intermittent score lines of varying lengths, which end in a fleck of aluminum debris. They can be easily identified by their chemical composition, which is the same as that of the extruded aluminum alloy [23]. The exogenous deposited layers were composed by a different material and they can be recognized as dye pick-up (Figure 5b—point A, Figure 5a—point A), which are particles worn-out of the extrusion dye and deposited on the semi-product surface [23–27].

Thereafter, the products, subjected to the whole production cycle, including painting and chromium plating, were also analyzed. The typical section, recorded via SEM-BSE, of the surface defects after these processes, is shown in Figure 6. It is worth highlighting that the paint layer does not follow the defect profile; it would thus cover and hide the defects featured by heights up to 15 μm.

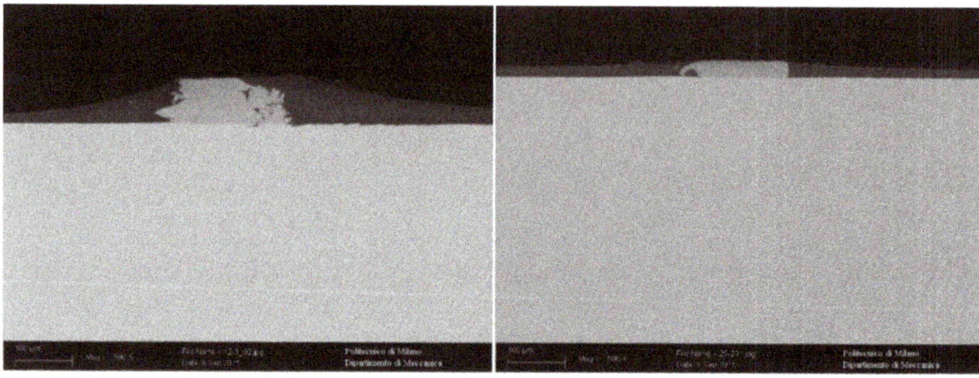

(a) (b)

Figure 6. SEM-BSE images of sections of the final products defects: sample 12 (**a**), sample 25 (**b**).

Through optical microscopy, a metallographic analysis was performed on the tested samples. Figure 7 sets out two micrographs obtained after threating the samples via Barker electrochemical etching. However, this analysis did not highlight any detail about the microstructure, which could be relevant or have influence on the generation of defects. Indeed, at the grain boundaries no secondary phase or trace of segregation phenomena is present [28].

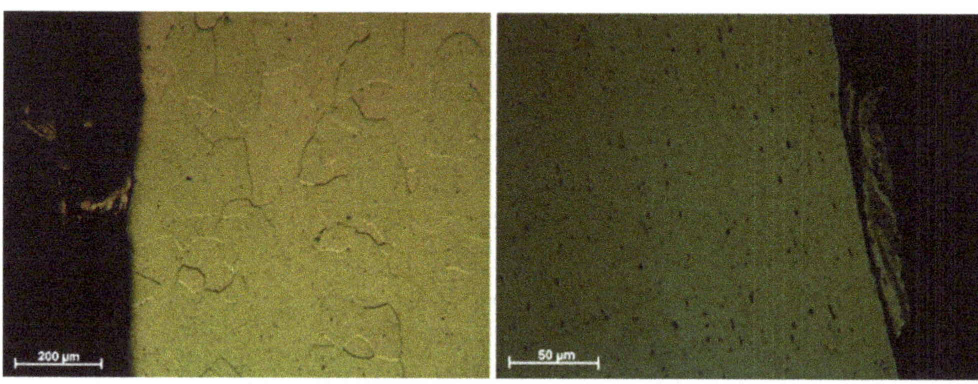

(a) (b)

Figure 7. Micrographic analysis of the pick-ups presents within the defect area the specimen 5 (**a**) and specimen 33 (**b**).

Moreover, the microstructures appeared homogeneous across the junction area sections (Figure 8a). Thus, no relationship between the microstructure and the surface defects was identified. Moreover, no differences in the morphologies of the crystal grains were detected; all the junction areas of the tested specimens featured coarse equiaxial microstructures, without been affected by any variations in the working parameters.

However, few millimeters away from the junction areas, the microstructures of the specimens were very different (Figure 8b). The grains displayed a lamellar structure, oriented along the extrusion direction, usually featuring the extruded semi-products. This occurred because the process arrest leaves the junction area for a much longer time in contact with the mold after being deformed. Thus, a recrystallization phenomenon was promoted, as testified by the loss of the extrusion deformation pattern of the grains [29].

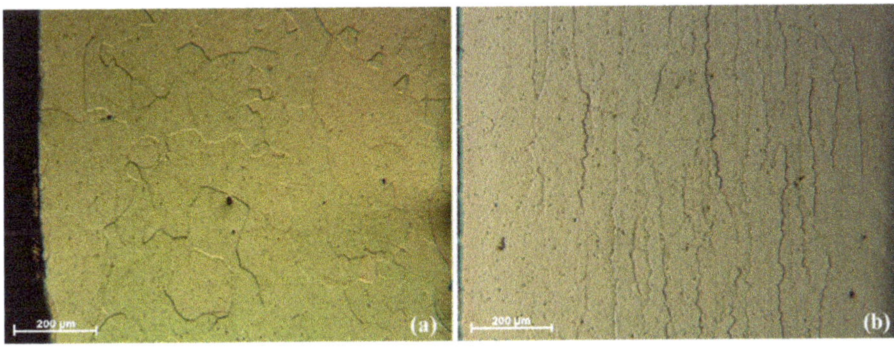

Figure 8. Micrographic analysis of sample 9 in the junction area (**a**) and in a different area representative of the rest of the semi-product (**b**).

Moreover, the grain size number G, an index measuring the average grain dimensions, was calculated through the Heyn intercept method and the results are reported in Table 3.

Table 3. Grain size number (G) index.

Sample	Extrusion Mean Temperature (°C)	G (N_L/mm)
5	526	14
9	534	13.5
13	545	14.5
21	528	14
25	528	14
33	511	14
37	506	14

Furthermore, a SEM-SE morphological analysis wasperformed, to qualitatively assess the differences between the different mold cooling methodologies. The most interesting results concerned billets extruded with the same working parameters. In Figure 9, the junction areas of billets 5-2 (air cooled dye) and 25-4 (liquid nitrogen cooled die) are displayed in detail. Compared to the 25-4 sample, the 5-2 specimen showed much larger defects and a large quantity of the material adhered onto the surface.

Figure 9. SEM-SE morphological analysis of the junction areas of the billets comparing one produced through an air cooled die 5-2 (**a**) and one produced via the liquid nitrogen cooled die 25-4 (**b**).

The results of the roughness tests, performed on the surface of the body of the extruded semi-product—hence not in the junction areas—are shown in Table 4. The mean roughness grew with the mold cooling liquid nitrogen valve aperture during the extrusion trials. On the other hand, the measured maximum roughness displayed a decreasing trend under the same operative conditions.

Table 4. Roughness tests results (R_a: Roughness Average; R_z: Average Maximum Height of the Profile; R_{max}: Maximum Roughness Depth).

Sample	R_a (μm)	R_z (μm)	R_{max} (μm)
13	0.40	3.02	4.31
25	0.44	4	5.12
33	0.53	3.54	3.89

The surface texture data, collected via microscopy focus-variation technology, are set out below—both the parameters describing the roughness (Table 5) and the height histogram of the surfaces and its statistics (Figure 10 and Table 6). The results are similar to those for the profile roughness. The mean values of the surface roughness slightly increased with use of the liquid nitrogen cooling system. On the other hand, the parameters describing the maxima and the minima of the surface, underwent a decreasing trend with the use of the liquid nitrogen cooling system.

In detail, the skewness values indicated that the height distribution of the liquid nitrogen-cooled sample was more "valley-tailed" than the traditionally extruded surface. For both distributions, the kurtosis parameter highlighted a leptokurtic behavior with a positive excess kurtosis. The higher value of the sample traditionally extruded testifies the presence of larger tails in its height distribution, due to the occurrence of larger surface defects in the sample produced through this technology, as observed via other characterization techniques as well.

Table 5. Surface roughness parameters.

Parameter	Description	(u)	Sample	
			5	37
Sa	Average height of selected area	nm	361	471
Sq	Root-mean-square height of selected area	nm	467	587
Sp	Maximum peak height of selected area	μm	1.99	1.41
Sv	Maximum valley depth of selected area	μm	3.98	2.48
Sz	Maximum height of selected area	μm	5.97	3.89
S10z	Ten-point height of selected area	μm	4.66	3.80
Ssk	Skewness of selected area		−0.45	−0.78
Sku	Kurtosis of selected area		3.69	3.13
Sdq	Root mean square gradient		0.08	0.09
Sdr	Developed interfacial area ratio	%	0.34	0.44
FLTt	Flatness using least squares reference plane	μm	5.97	3.89

Table 6. Statistics of the height histograms of specimens' surfaces.

Height Histogram Statistics	(u)	Sample	
		5	37
Number of Elements		8,816,414	8,253,132
Classes		120	195
Class Width	μm	0.05	0.02
Mean Value	μm	0.0186	−0.0225
Standard Deviation	μm	0.4667	0.5870
Minimum Value	μm	−3.9756	−2.4777
Maximum Value	μm	2.0244	1.4233

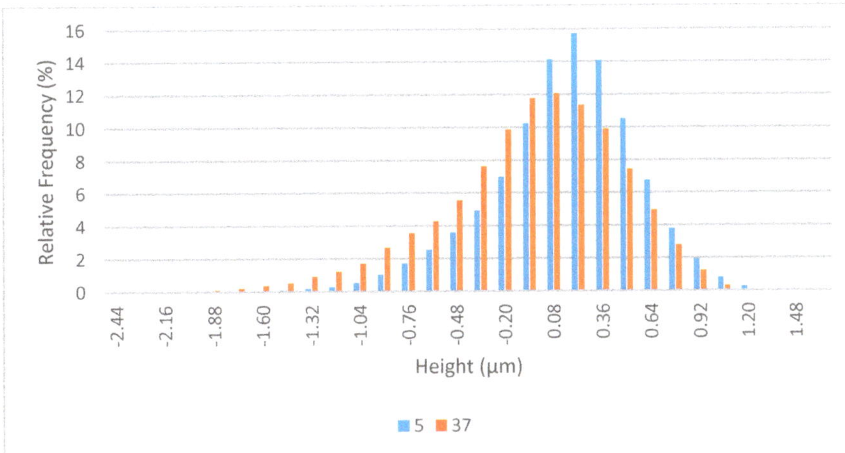

Figure 10. Height histogram of the samples' surfaces: traditionally extruded (specimen 5) and the liquid nitrogen cooled (specimen 37).

4. Discussion

The experimental data from the current study testify that the liquid nitrogen cooling effect is not instantaneous and a transition time is required before a steady working regime temperature can be reached and maintained. This is related to the mold's thermal inertia. Furthermore, when using this technology, the working temperature becomes lower relative to conventional air cooling of the extrusion mold. However, the differences between temperatures measured at the different dye holes broaden. This phenomenon may be related to the inhomogeneity of the heat removal, linked to the non-optimal sizing of the cooling channel. Therefore, this aspect should be taken into account during the design of the mold, to achieve easy and uniform temperature control [30].

The experimental data regarding the defects analysis in the junction area of the semi-finished extruded billets, found the presence of two varieties of defects. The first featured aluminum flecks of debris adherent to the surface. The chemical composition confirmed that the origin of this debris was the extruded material, since it is identical. The adhesion between the dye and the extruded material generated these defects, as testified by their peculiar morphologies. These defects can thus be recognized as pick-ups [23].

Further to this, a second category of defects was detected and it identified as dye pick-up. This observation was proven by the chemical composition analysis of the deposited material. Indeed, this composition is similar the composition of the AISI H13 tool steel, which is reported in Table 7. Since the extrusion dye is made in this material, its provenance is clearly proved [23–27].

Moreover, any metallurgical origin of the analyzed defects can be excluded. Indeed, both the SEM/BSE analysis and optical microscopy did not detect the presence of any other phases or intermetallic precipitates, any abnormal microstructural features or any abnormal chemical element concentrations. It follows that the origin of the detected defects is linked to wear issues and, in detail, it can be recognized as relating totally to the processing parameters and the tools [23–27].

Table 7. Extrusion dye material (AISI H13 tool steel) chemical composition, as designed by standards, the element concentrations are expressed as wt %.

Steel	% C	% Si	% Mn	% Ni	% V	% Cr	% Mo	% Fe
AISI H13	0.40	1.0	0.4	0.3	1.0	5.0	1.5	Bal.

It should be highlighted that no correlation between the variation of extrusion parameters during the production trials and the species defects were identified. The generation of these kinds of defects would not be totally prevented using liquid nitrogen cooling of the extrusion dye. On the other hand, the occurrence of new defect classes can be excluded by using this technology.

In addition, the liquid nitrogen cooling of the extrusion dyes resulted in a beneficial effect on the amount and on the size of the extrusion defects with patent improvements. Indeed, although the analyzed extrusion defects were few and topologically isolated, the use of this technology further lowered their occurrence and dimensions [11,12].

However, considering their application, in which the aesthetic aspect plays a predominant role, even isolated and very small defects would cause the rejection of the extruded semi-product (Figure 11). Taking into account the further steps involved in the realization of the finished product, we can set the defect acceptance threshold at a height of 15 μm. Indeed, since the paint layer does not follow the defect profile, it could cover and hide these small defects.

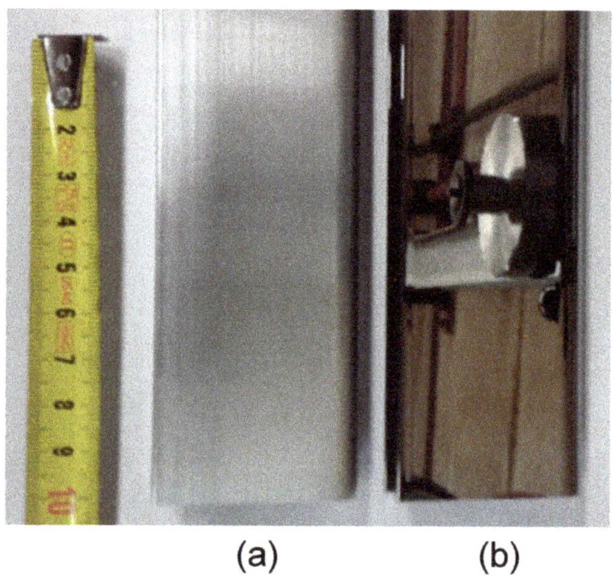

Figure 11. Semi-finished (**a**) and finished (**b**) products: the production cycle, after extrusion, is completed by brushing, acid degreasing, painting and chromium plating.

Considering this finishing process, the most significant roughness data were those describing the maxima of the roughness measurements of the surfaces and the distribution of the height histogram. These data (Tables 4–6 and Figure 10) demonstrate the beneficial effect of the liquid nitrogen cooling of the extrusion dye, and an improvement in this key aspect can be appreciated. In addition, the surface roughness maps of the reconstructed surfaces likewise highlight this aspect (Figure 12). The surface of the liquid nitrogen cooled specimen was very smooth and featured valleys along the extrusion direction. On the other hand, on the surface of the conventionally extruded semi-product, peak-like defects were detected emerging from the surface, harming the subsequent finishing processes.

Liquid nitrogen dye cooling technology improves the extrusion process through a complex mechanism, which can be described as follows: the debris, which generates the previously analyzed defects, is in both cases related to the adhesion phenomena occurring during extrusion between the dye and the flowing aluminum. However, this debris is much less present when the liquid nitrogen cools down the mold, even if the measured temperature is the same as for the air cooling.

Hence, the improvement in the surface finishing, obtained through the liquid nitrogen cooling technology, is not only related to the lowering of the working temperature.

A different kind of contribution, which could increase the surface finishing, is given by the mechanical action that the nitrogen exerts on the flowing extruded material. Indeed, while absorbing the heat cooling the dye, nitrogen undergoes a phase transformation from the liquid to the gaseous state. This phase transition is known to be accompanied by a very large increase in volume. However, since the cooling circuit is pressured, the phase transformation generates a detaching pressure on the flowing material, which keeps it separated from the mold.

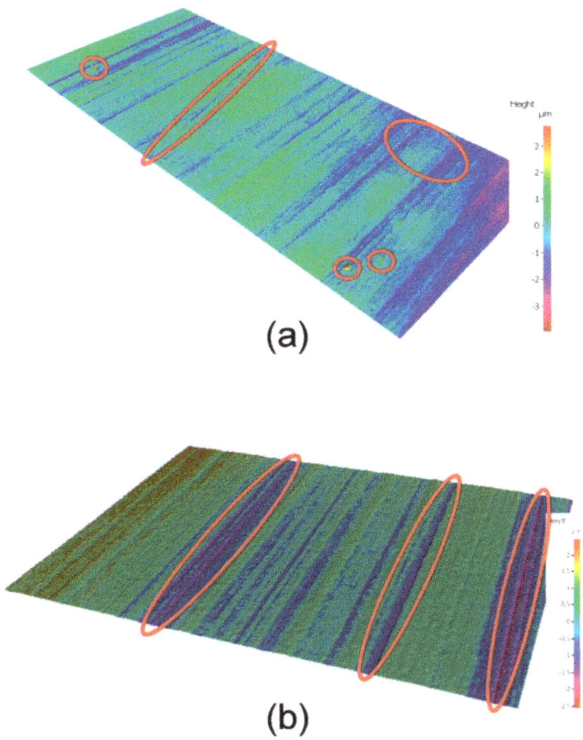

Figure 12. Surface roughness maps. The surface of the conventionally extruded semi-product (**a**) is featured by peak-like defects emerging from the surface. The surface of the liquid nitrogen cooled specimen (**b**) features the presence of valleys along the extrusion direction.

Moreover, liquid nitrogen cooling has another beneficial effect for surface finishing. In detail, it acts as an inert atmosphere enveloping the flowing material, avoiding its oxidation. This plays a central role in wear phenomena, in which aluminum is involved, since its oxide is very easily generated and is extremely hard. On the other hand, in comparison, aluminum is extremely soft and easily damageable with respect to its oxide [4].

Therefore, these beneficial effects on surface finishing, related to the use of liquid nitrogen mold cooling, result from a combination of the cooling effect, the inertizing effect and a mechanical detaching pressure generated by the liquid-to-gaseous transition.

5. Conclusions

Production trials of aluminum alloy extruded semi-products were performed both with conventional air cooling of the extrusion mold and with newly installed liquid nitrogen cooling of the mold. The semi-products analysis led to the identification of two main categories of defects found in the joint areas: pick-up and dye pick-up.

The metallographic analysis pointed out the absence of any hot shortness, grain boundary segregation or coarse intermetallic compound formation phenomena occurring within aluminum semi-finished products as concurrent causes for pick-up generation. The liquid nitrogen dye-cooling did not have any influence on the extruded aluminum grain size. However, the use of liquid nitrogen cooling reduced the adhesion phenomena between the flowing aluminum and the mold, hindering the generation of the detected defects. Moreover, liquid nitrogen dye cooling provided inertizing, detaching and cooling effects. Consequently, the best results, in terms of absence of defects, were achieved at the lowest working temperatures through the use of liquid nitrogen cooling. Furthermore, low working speeds also hindered the generation of defects.

Finally, liquid nitrogen mold cooling, compared to classic air cooling, granted smoother surfaces featured by a high density of valleys along the extrusion direction and the almost complete absence of peak-like defects emerging from the surface which would lead to flaws in the subsequent finishing processes.

Author Contributions: A.F.C., S.B., C.D.C., A.G., L.M. took part to the experimental campaign relative to the metallographic investigation and the defects detection and analysis. C.M. is the research group leader and supervised the work, G.M. and M.B. performed the extrusion an collect the process data using the software produced by A.t.i.e. uno.

Funding: This research received no external funding.

Acknowledgments: This research did not receive any specific grant from funding agencies in the public, commercial, or not-for-profit sectors.

Conflicts of Interest: The authors declare no conflicts of interest.

References

1. Sheppard, T. *Extrusion of Aluminum Alloys*; Kluwer Academic Publishers: Norwell, MA, USA, 1999.
2. Saha, P.K. *Aluminum Extrusion Technology*; ASM International: Materials Park, OH, USA, 2000.
3. Donati, L.; Segatori, A.; Reggiani, B.; Tomesani, L.; Bevilacqua Fazzini, P.A. Effect of liquid nitrogen die cooling on extrusion process conditions. *Key Eng. Mater.* **2012**, *491*, 215–222. [CrossRef]
4. Ko, D.-H.; Kang, B.-H.; Ko, D.-C.; Kim, B.-M. Improvement of mechanical properties of Al6061 extrudate by die cooling with N_2 gas during hot extrusion. *J. Mater. Process. Technol.* **2013**, *27*, 153–161. [CrossRef]
5. Bjork, T.; Bergstrom, J.; Hogmark, S. Tribological simulation of aluminum hot extrusion. *Wear* **1999**, *224*, 216–225. [CrossRef]
6. Bastani, A.F.; Aukrust, T.; Brandal, S. Optimisation of flow balance and isothermal extrusion of aluminum using finite-element simulations. *J. Mater. Process. Technol.* **2011**, *211*, 650–667. [CrossRef]
7. Ma, X.; De Rooij, M.B.; Schipper, D.J. Modelling of contact and friction in aluminum extrusion. *Tribol. Int.* **2010**, *43*, 1138–1144. [CrossRef]
8. Ma, X.; De Rooij, M.B.; Schipper, D.J. Friction conditions in the bearing area of an aluminum extrusion process. *Wear* **2012**, *279*, 1–8. [CrossRef]
9. Saha, P.K. Thermodynamics and tribology in aluminum extrusion. *Wear* **1998**, *218*, 179–190. [CrossRef]
10. Schikorra, M.; Donati, L.; Tomesani, L.; Kleiner, M. The role of friction in the extrusion of AA6060 aluminum alloy, process analysis and monitoring. *J. Mater. Process. Technol.* **2007**, *191*, 288–292. [CrossRef]
11. Celani, P. Liquid nitrogen technology. In Proceedings of the 8th Aluminum Two Thousand International Congress, Milano, Italy, 14–18 May 2013.
12. Ward, T.J.; Kelly, R.M.; Jones, G.A.; Heffron, J.F. The effects of nitrogen—Liquid and gaseous—On aluminum extrusion productivity. *JOM* **1984**, *36*, 29–33. [CrossRef]
13. Twigg, R.; Oesterreich, R. Extrud-AlTM—An alternative approach to nitrogen die cooling. In Proceedings of the 8th International Extrusion Technology Seminar, Orlando, FL, USA, 18–21 May 2004.
14. Cotell, C.M.; Sprague, J.A.; Smidt, F.A., Jr. *ASM Handbook Vol. 5: Surface Engineering. Surface Engineering of Aluminum and Aluminum Alloys*; ASM International: Materials Park, OH, USA, 1994; pp. 784–804.

15. Gali, O.A.; Shafiei, M.; Hunter, J.A.; Riahi, A.R. The influence of work roll roughness on the surface/near-surface microstructure evolution of hot rolled aluminum—Magnesium alloys. *J. Mater. Process. Technol.* **2016**, *237*, 331–341. [CrossRef]
16. Rao, K.S.; Rao, K.P. Pitting Corrosion of Heat-treatable Aluminum Alloys and Welds: A Review. *Trans. Indian Inst. Met.* **2004**, *57*, 593–610.
17. Rodríguez-Diaz, R.A.; Uruchurtu-Chavarín, J.; Valdez-Villegas, S.; Valdez, S.; Juárez-Islas, J.A. Corrosion Behavior of AlMgSi Alloy in Aqueous Saline Solution. *Int. J. Electrochem. Sci.* **2015**, *10*, 1792–1808.
18. De Rooij, M.B.; Ma, X.; Den Bakker, A.J.; Werkhoven, R.J. Surface quality prediction in aluminium extrusion. *Key Eng. Mater.* **2012**, *491*, 27–34. [CrossRef]
19. Li, L.; Zhou, J.; Duszczyk, J. Prediction of temperature evolution during the extrusion of 7075 aluminium alloy at various ram speeds by means of 3D FEM simulation. *J. Mater. Process. Technol.* **2004**, *145*, 360–370. [CrossRef]
20. Miller, W.S.; Zhuang, L.; Bottema, J.; Wittebrood, A.J.; De Smet, P.; Haszler, A.; Vieregge, A. Recent development in aluminium alloys for the automotive industry. *Mater. Sci. Eng. A* **2000**, *280*, 37–49. [CrossRef]
21. Berndt, N.; Frint, P.; Wagner, M.F.-X. Influence of Extrusion Temperature on the Aging Behavior and Mechanical Properties of an AA6060 Aluminum Alloy. *Metals* **2018**, *8*, 51. [CrossRef]
22. Tekkaya, A.E.; Schikorra, M.; Becker, D.; Biermann, D.; Hammer, N.; Pantke, K. Hot profile extrusion of AA-6060 aluminum chips. *J. Mater. Process. Technol.* **2009**, *209*, 3343–3350. [CrossRef]
23. Garbacz, P.; Giesko, T.; Mazurkiewicz, A. Inspection method of aluminum extrusion process. *Arch. Civ. Mech. Eng.* **2015**, *15*, 631–638. [CrossRef]
24. Björk, T.; Westergård, R.; Hogmark, S. Wear of surface treated dies for aluminum extrusion—A case study. *Wear* **2001**, *249*, 316–323. [CrossRef]
25. Birol, Y. Analysis of wear of a gas nitrided H13 tool steel die in aluminum extrusion. *Eng. Fail. Anal.* **2012**, *26*, 203–210. [CrossRef]
26. Qamar, S.Z.; Arif, A.F.M.; Sheikh, A.K. Analysis of Product Defects in a Typical Aluminum Extrusion Facility. *Mater. Manuf. Process.* **2004**, *19*, 391–405. [CrossRef]
27. Arif, A.F.M.; Sheikh, A.K.; Qamar, S.Z.; Raza, M.K.; Al-Fuhaid, K.M. Product defects in aluminum extrusion and their impact on operational cost. In Proceedings of the 6th Saudi Engineering Conference, KFUPM, Dhahran, Saudi Arabia, 14–17 December 2002; pp. 137–154.
28. Wang, L.; Cai, J.; Zhou, J.; Duszczyk, J. Characteristics of the Friction between Aluminium and Steel at Elevated Temperatures During Ball-on-Disc Tests. *Tribol. Lett.* **2009**, *36*, 183–190. [CrossRef]
29. Kayser, T.; Klusemann, B.; Lambers, H.; Maier, H.J.; Svendsen, B. Characterization of grain microstructure development in the aluminum alloy EN AW-6060 during extrusion. *Mater. Sci. Eng. A* **2010**, *527*, 6568–6573. [CrossRef]
30. Tercelj, M.; Kugler, G.; Turk, R.; Cvahte, P.; Fajfar, P. Measurement of temperature on the bearing surface of an industrial die and assessment of the heat transfer coefficient in hot extrusion of aluminum: A case study. *Int. J. Veh. Des.* **2005**, *39*, 93–109. [CrossRef]

MDPI
St. Alban-Anlage 66
4052 Basel
Switzerland
Tel. +41 61 683 77 34
Fax +41 61 302 89 18
www.mdpi.com

Metals Editorial Office
E-mail: metals@mdpi.com
www.mdpi.com/journal/metals

www.ingramcontent.com/pod-product-compliance
Lightning Source LLC
Chambersburg PA
CBHW062052220926
51850CB00010B/2741